Statistical mapping and the presentation of statistics

To my wife

Statistical mapping and the presentation of statistics

Second edition

G C Dickinson
Lecturer in Geography, University of Leeds

 **Edward Arnold**

© G C Dickinson 1973

First edition published 1963
Reprinted 1964
Reprinted 1967
by Edward Arnold (Publishers) Ltd.
25 Hill Street
London W1X 8LL

ISBN: 0 7131 5641 4

Photoset by The Universities Press, Belfast.
Printed in Great Britain by J. W. Arrowsmith Ltd., Bristol.

Contents

Acknowledgements

The author and publisher wish to express their thanks to Mr. P. Shrives and Mr. G. Bryant for their assistance with the illustrations in this book and to the Editors of *The Town Planning Review* for permission to use Fig. 2 from Vol. XXXIII, No. 1 (April 1962); to the Research Paper Series, Department of Geography, the University of Chicago for Fig. X-8 from G. F. White (ed.), *Papers on Flood Problems* (copyright by the University of Chicago); to *The Guardian* for a graph from the edition of 6th June 1961; to W. and A. K. Johnston and G. W. Bacon Ltd. for permission to use part of p. 13 of Bacon, *All Essentials School Atlas*; to the Army Map Service, Washington, for the figure on p. 23 of *A.M.S. Bulletin*, No. 11, September 1944; to the *Evening News* for permission to reproduce two posters; to *The Cartographic Journal* for permission to reproduce two computer maps from their December 1969 issue; to the University of London Press for permission to use the computer map on p. 34 of Kenneth E. Rosing and Peter A. Wood, *Character of a Conurbation*; and to the Controller of H.M. Stationery Office for permission to reproduce examples of a colmap and a gridmap from *Linmap and Colmap*, and a dotmap from the map library at the Department of the Environment.

Introduction

It was in 1753 that Mr Thornton, Member of Parliament for the City of York, opposed a Bill proposing to establish a national census for Great Britain by declaring that he did not believe 'that there was any set of men, or, indeed, any individual of the human species so presumptuous and so abandoned as to make the proposal we have just heard. . . . I hold this project to be totally subversive of the last remnants of English liberty . . . the new Bill will direct the imposition of new taxes, and, indeed, the addition of a very few words will make it the most effectual engine of rapacity and oppression that was ever used against an injured people.'

One cannot help wondering whether the citizens of York should not have been congratulated on the unusual perspicacity of their Member, for he and his fellow supporters (they were successful for a time) were, of course, attempting to stem not merely a single new Bill in parliament but something of much greater importance—the onset of a new era of government. Today, for better or worse, we live in that era which is among other things an age of measurement and statistics. We count, measure and calculate about all aspects of life in a manner which even Mr Thornton could scarcely have foreseen and our modern 'experts' are no longer revered personages drawing on long acquaintance in a particular field or working by intuition and 'experience', but men who can adapt and use this calculating most successfully to promote their own activities and predict the future.

Parallel with this pervasion of statistics into everyday life has come the need to acquaint an ever-increasing number of people with their implications. Governments are perhaps the most prolific 'counters' of all, and since official statistics can be a two-edged weapon it becomes important to ensure that the electorate understands both the nature and implications of such statistics relating to Government activity as may come to their notice. Growing awareness of the importance of public relations has led industry to follow a similar line, supplementing its long history of private calculating by placing before the public eye those facts and figures which it considers of interest or advertisement value; in fields such as Town and Country Planning it becomes essential to use all manner of statistics to summarise the present and predict future trends.

On all sides and through a wide variety of media we are constantly assailed by statistics—how our taxes or rates are spent, how many houses are needed, have been or will be built, how many more cars, or television sets or tins of meat or jam-jars made by Bloggs & Co. are used in preference to those made by their competitors, and so on; since it has long been recognised that 'dry' statistics are made both more intelligible and more interesting by pictorial or semi-pictorial representation there has also been considerably increased interest in the techniques available for this purpose. In many cases today where statistics are presented publicly it is the man in the street, the very general observer, whose attention is being sought and who is now taking his place alongside older-established, more specialised assessors of statistical material. In many fields of study, such as geography, economics and social studies it has long been impossible to present a complete picture without recourse to statistics and an understanding of them, and above all a knowledge of the techniques available for passing on to other people an idea of their implications, has always formed a fundamental requirement of teaching and research in those subjects.

It is to all those faced with this problem of presenting statistical information to a wider public that this book is addressed, and particularly to those who are encountering this problem for the first time, whether in the formulation of a thesis or piece of research, as part of a carefully organised publicity campaign or simply as a formal teaching requirement in a standard course of cartography at school or university. Essentially, this book aims to induce not merely an understanding of a few simple techniques but a new *approach* to the whole problem of statistical presentation. Properly used, the statistical map or diagram is no mere decoration added as an afterthought to improve the appearance of a piece of work. It must be treated as a highly specialised, often expensive (in time and labour) device which above all is an *integral* part of the task in hand, doing the job it was given to do and doing it not merely satisfactorily but as effectively as possible.

The author's experience in the course of several years' teaching of cartographic techniques to university undergraduates has shown that it is the need for this correct approach which is the most neglected aspect of statistical presentation, so that students frequently show too facile an appraisal of the task in hand, repeatedly using methods so conventional that they have almost become clichés, rather than seeking individual solutions to individual problems. In this they can hardly be blamed. The publications of their elders and betters are by no means immune from the same defect, and existing literature scarcely deals adequately with this need for the inculcation of a correct approach; it is hoped that this book will go some way towards remedying this situation.

Unfortunately for the less specialised person, we are passing today into a period of more intensive usage of statistics than ever before and one which is characterised especially by greater application of mathematical techniques to the figures themselves. In short we are witnessing greatly increased use of those techniques which the mathematician himself would consider part of that branch of his subject which *he* would term 'Statistics'. As its title implies this book is concerned with the *presentation* of statistics and will not consider what might be termed the 'mathematics of statistics'. Two good reasons can be advanced in support of this attitude. In the first place much useful work can be done with statistics, using no more than elementary mathematical knowledge, and it is the author's experience that the inclusion of even a small amount of relatively simple mathematics immediately deters a great many who would otherwise come seeking information. Whilst it cannot be denied that the all-inclusive approach is superior it is dearly bought if it discourages potential users from seeking instruction in that wide variety of techniques which involve nothing more complicated mathematically than long multiplication and long division. The second reason is a more fundamental one. The great majority of advanced mathematical techniques which can be applied to statistics seek to reduce their bulk and determine salient points and trends from within a mass of figures, but even when this simplification has been achieved, the results are not necessarily easy to appreciate and there will frequently arise the need to present them in a more graphic and interesting manner. No amount of advanced mathematics can over-ride the need for instruction in and appreciation of the techniques for *presenting* statistics as well as for *analysing* them, and since the mathematical aspect of statistics is plentifully and well written-up and that of statistical presentation is not, it is on this latter aspect of the problem that the attention of this book is concentrated.

Throughout this book the author has drawn consistently on his own experience, particularly in teaching cartography to undergraduates. The various defects in techniques and mistakes in approach which are described are those which experience has shown to be particularly prone to occur. In so far as it ventures outside the strict considerations of statistical techniques into subjects such as mathematics or source materials this book does so in an attempt to provide information, often of a remarkably elementary nature, which is nevertheless not obtainable easily and quickly from published works in those fields. Whereas the trained mathematician will unhesitatingly obtain the square root of 2543 from a table headed 'Square Roots 10 to 100' nine general readers out of ten would discard this and work out the answer more tediously, using logarithms. In like manner the chapter on source materials (a topic quite plentifully written up elsewhere) tries particularly to provide missing background information, the lack of which has been known to waste many hours of work and cause much frustration and disappointment.

It is not the aim of this book to be encyclopaedic in its nature or dogmatic in its approach. Taken individually, statistical techniques are as elementary in principle as they are complex in application, and consideration of both sides of their nature is essential before successful work can result. Acquaintance with the principle of the technique alone is of little use in itself, just as knowing how to make a motor-car go will not automatically produce a good driver. The statistical cartographer, like the experienced driver, must be prepared to adapt a relatively small store of essential 'know-how' flexibly and individually to the infinitely varied requirements of each situation which he meets. It is the aim of this book to help him to understand both of these aspects of the presentation of statistics.

1 Why draw statistical maps?

Anyone who has to prepare a statistical map or diagram might sensibly ask himself at the very outset, 'Why am I drawing this map?'[1] The question is by no means so unnecessary as it may appear at first sight. Of course, no one is likely to embark upon designing anything so complicated, tedious and often expensive, as the average statistical map without having a good reason for doing so, but in practice it is surprisingly easy for this initial aim to be forgotten as the map progresses and more pressing practical difficulties arise and call for a solution.

Many of these practical problems may occur when the designing of the map has scarcely got under way. A suitable technique must be chosen, compatible not only with the statistics themselves, but with the scale of the finished map, which in turn will often be determined by the space available to contain it; from the statistics may come difficulties, such as an excessively large range and inconsistency of definition or accuracy, whilst other problems of presentation may arise from the use of only a limited number of colours, or restriction to black and white only.

All of these are very real problems which must be solved before the designing of the map can proceed, but they should never be allowed to dominate the map so that its initial purpose is subjugated to their solution. It is not uncommon to encounter maps where this has happened; practical difficulties have come to the fore and detail so dominates the map that it becomes a mere formality of illustration in which one cannot 'see the wood for the trees', a poor substitute for the stimulating visual aid to understanding which a well-designed map ought to be.

There is no better way of counteracting this tendency than by starting out with the general aim of the map clearly in mind. The chapters which follow describe many more factors which must be accommodated within the map, but certainly no other single item will be so important in influencing the design as the basic reason for which it was drawn; since considerations of this kind are just as valid when reviewing the design of statistical maps generally as they are when constructing a single example, it is worth while examining, at the very outset, the reasons which cause people to design statistical maps.

Although many particular reasons might be associated with the need for any individual map, most of these will be found to fall within four broad categories which are sufficiently important to be defined first and elaborated afterwards. They are:

   i   to arouse greater interest in the subject matter presented

[1] To avoid tedious repetition the terms 'map' and 'statistical map' may be regarded as including 'diagram' or 'statistical diagram' throughout this book.

2   to clarify it, simplify it or explain its more important aspects
3   to prove a point referred to in text or speech
4   to act as a statistical 'quarry' for other users.

Of these four aims the first two are widely recognised and in practice almost inseparably linked. They have already been touched upon in the Introduction and need little elaboration. Few items are more unattractive to the eye or more tedious to read than long tables of figures, and unless one has a particularly good head for such things an attempt to read one of these tables may be disappointingly unrewarding. By the time the end of the table is reached the average reader is overwhelmed with information and quite unable easily to answer important questions, such as what were the largest and smallest quantities in the list, where or when did they occur and is there a tendency for groups of similar values or for any exceptional values to be present? The representation of this information in graphical form not only makes the whole subject appear more interesting, but also makes the answers to many of these questions easier to obtain.

The idea of using a statistical device to prove a point is so obvious that it is unlikely completely to be neglected. One does not consciously design a map which gives an entirely wrong impression of the subject under discussion, but it is very easy to be satisfied with too little, and every cartographer ought to ensure that the device used really does prove his point in the most convincing manner. The works manager who draws up a graph of increasing output to impress his board of directors will not choose so cramped a vertical scale that production has to increase by 25% before the line begins to climb at all noticeably when a more generous vertical scale would allow much smaller increases to convey a more forceful impression.[1]

Much the most neglected of these four main reasons is the idea of making the map a statistical quarry, providing a source of information for other purposes than those for which the map was originally designed. The basis of this idea lies in the capacity of the map to act as a remarkably concise summary, able to convey a very great deal of information, the description and implications of which could otherwise be explained only in many pages of text.

This must surely be one of the ideas which underlies the recent appearance of many specialised atlases devoted to individual countries or topics and in which the greater part of the information is displayed in map form. Unfortunately, in so many cases this aim appears to have been lost whilst the atlas was under construction and the mapping of the statistics has come to be regarded as the end point of the process; no provision has been made for the person who wants to 'read back' from the maps to the statistics themselves and it is impossible to adapt and use the material displayed for other research in parallel fields. Figure 1 illustrates a map which was designed with both this last-mentioned point and the general aim of acting as a statistical 'quarry' very much in mind. Not only is some form of diagrammatic representation used to summarise the main points of the distribution but the actual figures involved are given in shortened form alongside each symbol.

These then are the four great themes which form a basic framework within which the statistical cartographer must work and against which the ideas put forward in the rest of this book must be placed. It is almost self-evident that, expressed in interrogative form, they provide one of the most searching 'acid tests' with which to measure the success of any individual design.

[1] The prudent works manager will not stretch this point too far, of course; an over-generous vertical scale will tend to make even the slightest *fall* in production look alarming and the solution must lie in moderation— or a quickly redrawn version of the graph if production begins to fluctuate.

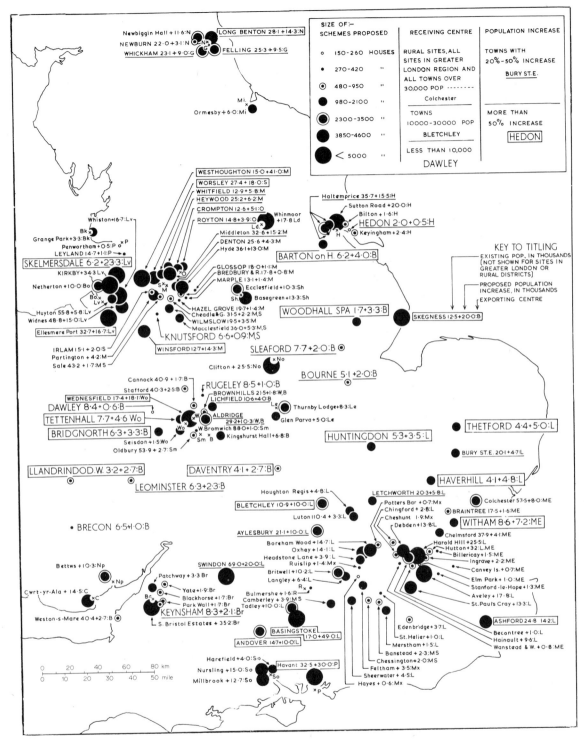

Figure 1   Proposals for the reception of overspill in England and Wales, Autumn, 1960

Key to exporting centres: B Birmingham: Bk Blackpool: Bo Bootle: Br Bristol: C Cardiff: G Gateshead: H Hull: L L.C.C. or London area: Ld Leeds: Le Leicester: Lv Liverpool: M Manchester: ME Metropolitan Essex: Mi Middlesbrough: MS Metropolitan Surrey: Mx Middlesex: N Newcastle: Np Newport: O Oldham: R Reading: S Salford: Sh Sheffield: Sm Smethwick: So Southampton: W Walsall: Wo Wolverhampton

This map was designed to provide, on sight, *detailed* information regarding these proposals, as well as to give some indication of the nature of their distribution. *Source:* Information supplied privately by Local Authorities

2 Statistical techniques

Once the *raison d'être* of any map has been clearly defined the cartographer is ready to turn to the more practical aspects of its construction. Amongst his first requirements will be a suitable method for presenting the material in question and this chapter attempts to review some of the many techniques which form the starting-point to the practical side of statistical cartography; a knowledge of their variety, the particular purposes to which each is best suited and the difficulties which their application involves is fundamental to the achievement of successful results.

Despite their number the great majority of these techniques are based on extremely simple ideas which are easily understood, but a full appreciation of the value of any one method demands much more than this, involving recognition of such features as its limitations and effectiveness in actual use and the (often) unintentional implications or 'slant' which it may convey to a particular set of figures. Unfortunately, these are factors which are not easily assessed and they may most satisfactorily be approached through consideration of certain basic principles and problems of map design which will be discussed in the later chapters of this book. This chapter will serve simply to introduce the techniques, describe their basic underlying ideas, explain any practical difficulties which may arise in drawing and using them, and consider some of their more individual characteristics.

Unlike the very considerable number of statistical techniques the basic types of statistical problem or distribution which they have to illustrate are relatively few. Moreover, since most of the methods are primarily designed to serve only one or two of these major distributions this offers a useful possibility of classifying their variety and reducing it to manageable proportions. Figure 2 illustrates such a classification and also suggests a logical order in which to comment on the individual methods. Naturally neither Figure 2 nor this book as a whole can hope to deal with all the various techniques which have been used to portray statistics, nor is it necessary that they should do so. By confining detailed description to the two dozen or so methods in common use they offer a reasonable guide to the devices most often encountered and at the same time provide sufficiently varied evidence on which to establish a set of principles and deductions which may guide the reader in assessing the effectiveness of any new method he may devise or encounter.

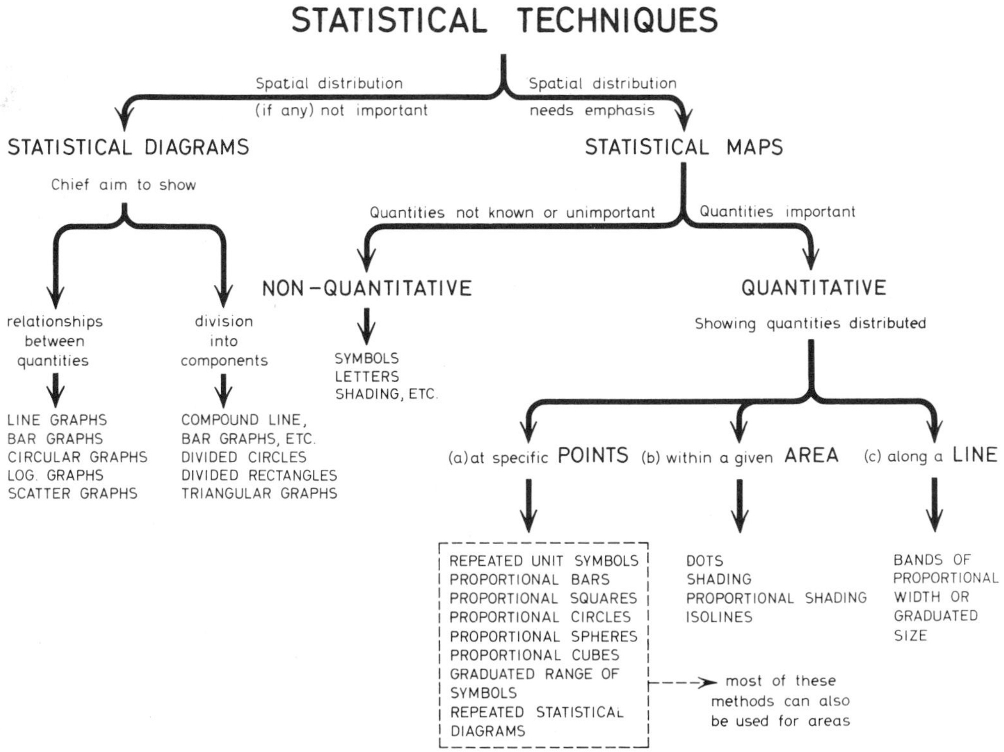

STATISTICAL TECHNIQUES

Figure 2   A 'family tree' of statistical techniques.

2.1   Statistical diagrams

The first broad subdivision of techniques suggested in Figure 2 is that into statistical diagrams and statistical maps proper. In the latter category the statistics which are being represented contain an element of spatial distribution which is either so important or so complicated that a base map is need on which this distribution can be recorded. In many cases, however, there is no need to do this; the statistics may refer to one place only or, where several localities are involved, their spatial distribution may be so well known that no representation of it is needed (e.g. in Figure 9A (p. 31), which shows the six major producers of copper in the world, it is unnecessary to invoke a method which actually shows where such places as Chile and Canada are).

Techniques falling into this second broad category will be referred to in this book as statistical diagrams. Strictly speaking this term should also include the many, often complicated, techniques which belong to that branch of mathematics known as statistics, but these are regarded as falling outside the scope of a work of this kind which, as explained in the Introduction, deals only with simple methods capable of being understood without mathematical training. The division into 'maps' and 'diagrams' is not entirely a rigid one. It will be seen later that statistical diagrams are often incorporated into techniques which have a map as their basis, but this slight degree of overlap need cause no confusion.

A TECHNIQUES SHOWING RELATIONSHIPS BETWEEN QUANTITIES

The various techniques comprised within the category of statistical diagrams can themselves be divided into two major groups, although this division is not a precise one. The distinction is basically between those methods which are concerned with showing the relationships between varying quantities and those which indicate the portion of the whole formed by several component parts.

In both divisions the various members of the graph family are well represented and since many methods in the first category can easily be adapted to perform the functions of the second they will be considered in their simpler form first and elaborated later.

Line graphs

Sooner or later most people encounter the simple line graph. Patients in hospital craning their necks to read a temperature chart and dripping holiday-makers hopefully examining the barograph on a rain-swept promenade are alike seeking reassurance from this elementary but effective device. The principle behind it is simple. Graphs show the way in which changes in one quantity—the patient's temperature or the height of the barometer—are related to changes in another quantity—in both these cases time. Mathematicians call these changing quantities *variables* and furthermore speak of one as the *dependent* and the other as the *independent* variable. The distinction is not difficult to understand. In most cases where simple graphs are used it will be noticed that one variable (the independent), e.g. time, changes steadily and regularly whereas the second (the dependent), e.g. temperature or pressure, changes more irregularly and may even be controlled by the first. It is usual when plotting information as a simple line graph to plot the independent variable on the horizontal scale, and the dependent variable on the vertical scale.

The basis of any graph is a series of values for the two variables plotted along two scales drawn at right angles; the name *line graph* is derived from the practice of emphasising the relationship between these plotted values by joining the dots either by a smoothly curved line which passes easily without arbitrary changes of direction through all plotted points (as in Figure 3A) or, alternatively, by a series of straight lines drawn between the plotted values (see Figure 3B). The line attracts attention to such features as maximum and minimum values and particularly to the tendency of values to rise or fall on any part of the graph. Normally the smooth curve is used only if the values plotted are part of a steadily changing relationship for which it might be desirable to determine intermediate values which were not actually recorded. Figure 3A shows such a graph illustrating the height of river level and the passage of a flood on one of the tributaries of the Tennessee River in 1875. A smooth curve was needed here between the plotted points to record the *gradual* rise, passing of the main flood wave and its recession, including the temporary slackening in the rate of recession which occurred between 3 and 6 March. Straight lines drawn between plotted points would have given sharp arbitrary peaks which could not have occurred in practice, and intermediate values which were either too high or too low according to whether the trend of the graph was concave or convex. The smoothly drawn curve allows accurate deduction of such information as the length of time during which the river level was above bank height (or any other particular height, e.g. the

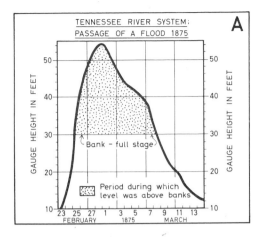

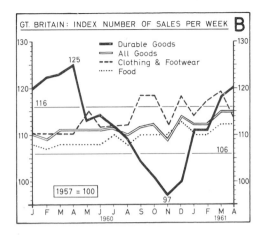

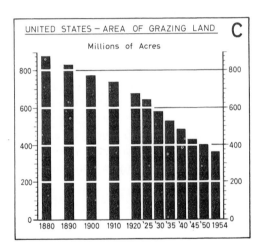

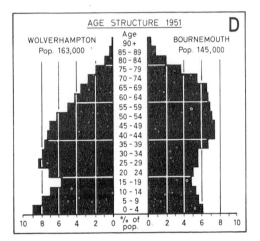

Figure 3 Types of simple line and bar graph.

Sources: **A**: G. F. White (ed.), *Papers on Flood Problems*, Chicago, 1961. **B**: *The Guardian*, 6 June 1961. **C**: U.S. Bureau of the Census, *Historical Statistics of the United States, Colonial Times to 1957*, Washington, D.C., 1960. **D**: Registrar-General, *Census of England and Wales, 1951*, County Volumes Staffordshire and Hampshire, London (H.M.S.O.), 1954.

height of any contemplated flood defences), even though the precise moments at which these levels were exceeded were not recorded in the series of values from which the graph was plotted.

Unless accurate interpolations of this kind are required from a graph the refinement of using a smooth curve to join up plotted values is not necessary. In many instances not only are intermediate values not required but it may be quite impossible for them to occur, as for example where figures refer to monthly or yearly totals; in these cases a straight line joining the plotted values is all that is required.

Although graphs are usually *constructed* on graph paper (commonly obtainable with squares of 1-mm, $\frac{1}{8}$-in, and $\frac{1}{10}$-in side) it is customary today to *publish* them on a blank background. In the past the 'graph paper background' was frequently reproduced as well, but the appearance was often rather heavy and fussy, detracting from the main message of the graph. Unfortunately, the contemporary trend represents a swing too much in the other direction, and it is not uncommon to encounter graphs on which the only positive indication of the values of the dependent variable (the quantity on the graph which arouses most interest) is a single, often badly calibrated, vertical scale at the extreme left-hand edge of the graph, a rather niggardly provision which makes it difficult for anyone to estimate values from the graph for his own purposes. If one dismisses as superfluous the old graph-paper lines which formerly carried values right across the face of the graph, the very least which can be attempted is to add a second vertical scale at the right-hand edge, so that a ruler placed across the graph from the two scales will easily allow values to be determined. It is also often useful to rule and mark two or three horizontal lines at values which are particularly relevant to the trends on the graph and to mark, in figures against the line, the values of any important maxima, minima or points at which the trends of the graph show important changes. On Figure 3B the values 106 and 116 are inserted to emphasise that most of the indexes shown fluctuate only between these two limits whilst the maximum (125) and the minimum (97) of the index for durable goods, the most important feature on the graph, are also marked. The aim is to present a more detailed portrayal of the situation than the rather bald message which derives from the shape of the lines alone.

It is occasionally possible to write on the face of the graph the value of every point from which the graph was constructed. Where this is done the value is often marked by a large dot whilst the lines between the dots can be drawn thick and heavy for clarity. This is a very bold style of graph which would normally be dismissed as too heavy for accurate reading, but where the values are added this point is no longer valid and the whole of the design of the graph can be orientated towards producing a forceful impression. Figure 5A (p. 22) illustrates such a graph.

Where the number of illustrations is restricted, or where comparisons are important, several items can be plotted on the same graph, though this practice can easily be abused. The more detail there is on any graph the more difficult will it be for the eye to single out the important features of any one line, and if the lines overlap at all the impression can become confused as can be seen on Figure 3B. Here, though individual lines can be traced, the characteristics of each are only obvious after fairly close examination. This is generally a disadvantage, but may not always be so; in Figure 3B itself the confused effect actually helps to put over the main message of the graph which seeks to show the way in which sales of durable goods have stood out distinctively from other main groups. Alternatively if the various quantities shown occur in different parts of the graph then a considerable number of lines can be accommodated without causing confusion.

It is not uncommon for two vertical scales to be needed on the same graph, e.g. where two quantities expressed in different units have to be recorded, as in Figure 4A. In this case the two vertical scales are quite independent and can be designed so that the two lines of the graph are close together or widely separated, whichever may be desired.

In Figure 4A the two vertical scales refer to quite different units—namely, square metres of cloth and metric tons of yarn, and this causes no confusion; generally speaking it is not desirable to incorporate on one graph two different vertical scales for the *same* units, however tempting this may be. The result will almost certainly produce a strongly

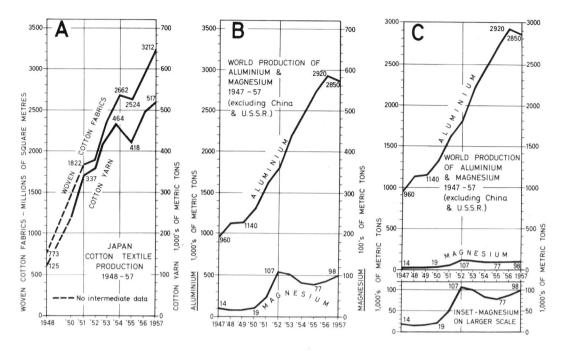

Figure 4   Aspects of Vertical Scales on Line Graphs. **A**: Two different vertical scales for two different units. **B**: Two different vertical scales for the same units. **C**: An improved method of presenting the information in B.

Source: United Nations, *Statistical Year Book, 1958,* New York, 1958.

misleading visual image which will not easily be totally overcome. Figure 4B shows recent world production of aluminium and magnesium, two metals frequently alloyed together and of common interest to modern technologists because of their lightness. Unfortunately, production of magnesium is much less than that of aluminium so that there is a strong temptation, if the two are graphed together, to use a different vertical scale for magnesium. As Figure 4B shows, even when significant values are added to the graph it is not easy to counteract the first impressions which one receives of the relative magnitude of production of the two metals in any one year. In 1952 magnesium production seems to be about one-third of that of aluminium, whereas in fact it was about one-twentieth. A better solution to the problem is shown in Figure 4C where the two are graphed comparably, but an 'inset' graph on a more generous vertical scale is also added for magnesium.

Whether one or two vertical scales are in use on a graph it is important that the vertical scale or scales should start at zero, if changes in values are to be seen in their true proportion. Unless this is done the graph can exaggerate small changes so that it is impossible to get any real idea of their significance. Occasionally the changes may be sufficiently important on their own to need graphing in this way or, as Figure 5A shows, there may be an ulterior motive in doing so, but in most cases a more sensible picture will be obtained by allowing the vertical scale to show the whole range of values from zero (see Figure 5B).[1]

[1] This example is not entirely fanciful. Figure 5A was suggested by an actual poster pushed through the author's letter-box during the course of an election campaign.

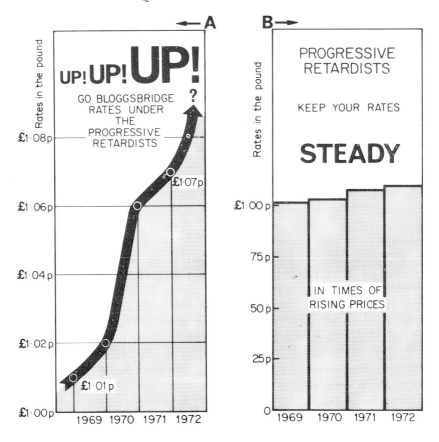

Figure 5 Two different ways of portraying the same information may produce two quite different impressions. **A** attempts to exaggerate the importance of changes, whereas **B** shows these in their true proportions. The designation 'steady' may not satisfy all ratepayers, but it at least presents the situation in rather a better light.

Bar graphs

Very similar indeed to the line graph is the *bar graph*, an example of which is shown in Figure 3C (p. 19). It originates in the same way from a series of values plotted against two axes, but in this case instead of being joined by a line the values are represented by a series of vertical bars which run from the plotted points to the foot of the vertical scale. Usually each bar is kept distinct from its neighbour, but this is not absolutely necessary.

The effect is rather similar to the line graph if only for the reason that the tops of the bars trace out rise and fall of the total, but on the other hand the presence of the bars themselves attracts attention to the actual *quantities*, whereas the line graph attracts the eye away from the quantities and towards the rise and fall of the values. Quite apart from this rather subtle, psychological distinction the bar graph has certain other advantages. It is particularly suitable where the data represent values (e.g. total annual output of any commodity or total rainfall in any one month) which are quite separate and distinct from the preceding and succeeding ones, the break between the bars emphasising this distinction. In Figure 3C the use of the bar graph technique also makes it easier to emphasise the fact that after 1920 the data changed from a 10- to a 5-yearly basis, than would have been the case with a line graph.

In very many cases, however, the bar graph and line graph are interchangeable and the kind of considerations which apply to such features as vertical scales, adding pertinent values, etc., in the line graph apply equally here. Thus on Figure 3C a line has been ruled across the bars every 200 million acres to help estimation of values and, incidentally, to emphasise the date when the amount of grazing land fell below these levels.

Although line graphs are almost invariably drawn using the vertical scale for the dependent variable, this rule is frequently ignored where bar graphs are concerned and these are often to be seen 'turned on their side', with horizontal bars, as in Figure 3D (p. 19). Not the least advantage of this practice is the ease with which such features as dates, names and values can be written in when the bars are horizontal, whereas the narrow vertical bar often allows this only if the printing is made to stand on end.

As with the line graph it is possible to show two or more items on the same bar graph. The usual procedure is to place the bars for the various items side by side, but it will be appreciated that if more than three variables are recorded each bar becomes separated from its predecessor and successor by rather a large gap so that it is not always easy for the eye to follow trends as well as values. This disadvantage can partly be overcome by slightly overlapping the bars, but even so this is one instance in which the line graph has a definite advantage over the bar graph.

Circular graphs

It may sometimes happen that a series of values which has to be plotted relates to a recurrent state of affairs. The most commonly encountered data of this kind are figures for mean monthly temperature and rainfall though other examples are not difficult to imagine; a large store will normally vary the proportion of its staff employed in different departments throughout the year and might prepare figures to show the average percentage of the staff employed in any given department each month throughout the year; similar data could relate to labour employed in the various operations on a farm.

Where cyclic features of this kind occur, their representation by normal line or bar graphs has certain disadvantages, notably that the left and right hand edges of the graph break what is manifestly a continuum and may make it difficult to appreciate trends at a critical time. It is surely a major defect that with climatic statistics relating to the northern hemisphere, and presented in the usual manner, the winter half of the régime can only be seen by piecing together two separate portions; in the southern hemisphere matters are worse, for it is the figures for the more important summer growing season which are interrupted.

It is a simple matter to overcome this objection by devising a circular form of the normal graph and the result has been variously called a circular graph, clock graph or polar chart. The basic idea is simply that of taking the normal graph and stretching and bending it so that the two outer edges meet. Values now increase radially outwards and it is usual to make zero a small circle since this makes drawing easier than if zero were placed at the

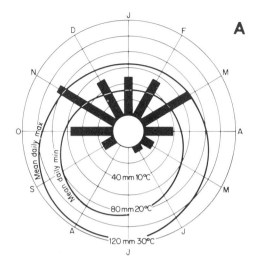

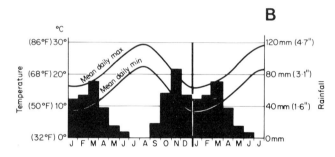

Figure 6 **A:** A circular graph or polar chart—mean monthly temperature and rainfall at Málaga, Spain. **B:** An alternative method of presenting the same information.

Source: Air Ministry, Meteorological Office; *Tables of Temperature, Relative Humidity and Precipitation for the World*, Part III, *Europe and the Atlantic Ocean*, London, H.M.S.O., 1958.

very centre of the circle. Figure 6 illustrates such a graph showing mean monthly temperature and rainfall figures for Málaga in Spain, a station which is particularly poorly served by normal methods because of the way in which they interrupt the representation of the winter rainfall with its peculiar double maximum.

Generally speaking, circular graphs are not widely used. Though they may overcome the disadvantages described above they bring additional handicaps which render them less attractive. The greatest of these is that the easily appreciated rise and fall of the normal graph is replaced by the idea of a line moving away from or towards the centre of the circle and considerable practice is needed before its message can be appreciated in detail. It is not easy in Figure 6A to form much of an impression about what is happening in the yearly temperature régime at Málaga, and even with the less difficult bar graphs for rainfall their divergent nature weakens the side-by-side type of comparison. On the whole the disadvantages of interrupting cyclic phenomena can be more easily overcome by graphing $1\frac{1}{2}$ (or even 2) cycles in the normal way, as in Figure 6B, although where this is done the end of each cycle should be clearly marked to avoid confusion.

The circular graph is the first example we have met of a feature which occurs more widely than it should in statistical techniques—namely, a method which is perfectly sound, even logical, mathematically, but which, visually, is rather ineffective.

Logarithmic graphs

So far we have considered graphs in which changes in *quantity* were all-important, i.e. where we wish to compare actual *amounts* of increase and decrease. This is not always the case. For example, an efficiency expert comparing changes in output at two factories after new methods of working have been introduced will not be interested simply in increase in the *amount* of production since one factory may be much bigger than the other and would naturally show a greater numerical increase. What is needed is some method of assessing the *rate* of increase rather than the amount; if output from the smaller factory has trebled and that from the larger only doubled, the new methods have obviously been more successful in the former whatever the quantities produced may seem to indicate.

The need to pay attention to *rate* of increase as well as amount is quite common and it is interesting to consider how this problem could be presented graphically. Unfortunately in studies of this kind the ordinary line graph is not only useless but sometimes positively misleading. Consider another example. At two villages X and Y the populations at three successive censuses were 100, 200 and 400 and 50, 100 and 200 respectively. In which village is the population increasing most rapidly? Graphed in the normal manner, as in Figure 7A, the impression is given that the population in X is increasing far faster than in Y, particularly so between 1921 and 1931. This is quite true, numerically, but a glance at the figures will soon indicate that in fact *both* villages have *consistently* shown the *same* rate of increase, each doubling its population every ten years. The line graph is not particularly successful here; what is needed is another sort of graph which will *show equal rates of change by lines of equal slope*, whatever the numerical basis of that change may be.

The logarithmic graph is such a device, and an example is shown in Figure 7D; though at first sight it appears a little strange it is not difficult to understand once the basic idea is grasped. So far as the horizontal scale is concerned all is quite normal;[1] it is in the vertical scale that changes are introduced. Here the numbers are not spaced evenly, e.g. in Figure 7B the interval between 20 and 30 is slightly less than that between 10 and 20 and the same amount separates 200 and 300 higher up the scale. This is because on the vertical scale of a logarithmic graph numbers are spaced not according to the difference between them but *according to the difference between their logarithms*. Figure 7C shows how this arrangement works and Figure 7D the kind of graph that results, characterised by a gradual and regular 'bunching' of lines on the vertical scale.

Each of these regular bunchings is called a *cycle* and it will be noticed that the top and bottom line of each cycle must be 10 or some multiple or decimal of 10, e.g. 100, 10,000, 1, 0·01. Each time a new cycle is added to the graph the value at the top is 10 times greater than that at the bottom and it follows from this that going down the scale, however many cycles are added, *zero can never be reached on a logarithmic graph*.

Although it is perfectly possible to draw out one's own vertical scale for a logarithmic

[1] In another type of logarithmic graph both vertical and horizontal scales are logarithmic, but such a device falls outside the scope of this book.

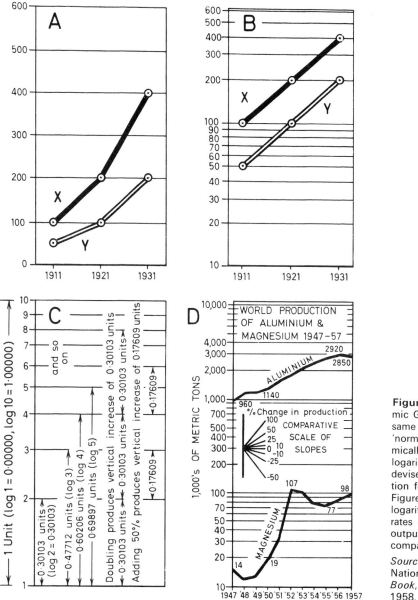

Figure 7 The Logarithmic Graph. **A** and **B**: The same information graphed 'normally' (A) and logarithmically (B) **C**: How a logarithmic graph is devised. **D**: The information formerly presented in Figures 4B and 4C graphed logarithmically to allow rates of change in annual output to be deduced and compared.

Source: Figure 7D. United Nations, *Statistical Year Book, 1958*; New York, 1958.

graph by spacing values according to their logarithms obtained from mathematical tables this is a tedious process and these graphs are usually plotted on specially ruled graph paper. Such paper is not of uniform character, for both the number of cycles on the vertical scale and the spacing of lines on the horizontal scale can vary and it is usually described by specifying these two quantities, e.g. 4 cycles × 1 mm or 3 cycles × $\frac{1}{10}$ in, the latter figure referring to spacing on the horizontal scale.

Figure 7D redraws logarithmically the figures for world production of aluminium and magnesium already encountered in Figure 4C (p. 21) and enables interesting comparisons to be made not between *amount* of production but of *rates of change* of production and *fluctuations*. In Figure 4C world production of magnesium misleadingly appeared rather steady; Figure 7D shows it to be subject to quite violent fluctuations regularly rising or falling by between 10% and 25% each year though showing a sustained increase of more than 100% per year between 1950 and 1952. On the other hand aluminium, with a more established position in world technology, shows a remarkably steady expansion at between 20 and 25% per annum for most of the period up to 1956. It is not difficult to add to a logarithmic graph a scale indicating the slopes produced by typical percentage increases and decreases and this has been done in Figure 7D to help estimation.

The use of logarithmic graph paper is sometimes advocated as one way of graphing figures which contain a large vertical range which would be difficult to accommodate on the normal line graph. The rapid rate of increase on the vertical scale of the logarithmic graph is very tempting here, but this use is not to be advocated; a logarithmic graph is a specialised device for fulfilling a particular need and should be used as such. It is not without its drawbacks—the man in the street is inclined to be highly suspicious of graphs with peculiar vertical scales and logarithmic graphs are not likely ever to be accepted popularly as line graphs are—but for performing its particular task it has no substitute and it might with advantage be rather more widely used than is the case at present.

Scatter graphs (scatter diagrams) [1]
The logarithmic graph is a mathematician's device which it is not too difficult for the layman to understand; the same can be said of the scatter graph. True, to obtain the finer implications of this device, or the most accurate deductions from it, considerable mathematical knowledge is needed, but it has nevertheless several simple uses and functions which the layman can understand quite easily, and these simpler aspects will be covered in this book.

The graphs which have been studied so far are used in determining the relationship between a series of *several* values for two variables related to *one* place, example or commodity. It is not difficult to envisage a slightly different kind of investigation where there is only *one* value for each of the two variables, but this is known for *several* places. Each value can still be plotted against scales drawn on two axes at right angles in the usual way, but in this case there is no question of joining the points by a line since no continuity is involved and each example is separate and distinct. All that we have, therefore, from which to read the message of this graph is a scatter of points, hence its common title of 'scatter graph'.

The scatter graph is used to investigate what sort of relationship, if any, exists between two variables which occur over a wide area. It could be used, for example, to see if there is any clear relationship between birth-rate and standard of living in the various countries of the world, total population and number of persons employed in service industry in a

---

[1] The term 'scatter diagram' is the one most commonly used to describe this device in textbooks on statistics. The term 'scatter graph' is used here to stress analogies in construction with ordinary graphs and in use with triangular graphs (see below).

group of towns, size of factory and output per worker in a group of factories in the same industry, and so on.

Any such relationship has to be deduced from the patterns which the scatter of dots shows when the values are plotted. If the dots show a random scatter, with no grouping or obvious trends, as in Figure 8B, then no systematic relationship exists between the two variables under investigation. On the other hand the dots may show a marked tendency to form groups, as in Figure 8C where two main groups may be distinguished. The very fact of the proximity of the plotted values on the graph indicates that here are places which are very similar in this respect, and the simplest possible use of the scatter graph is in defining groups of similar places without attempting to carry the investigation any further. The essential implication of Figure 8C is that the places under investigation tend to fall very markedly into two main groups. This might lead to further research to determine reasons for this grouping and also for the few examples on the graph which

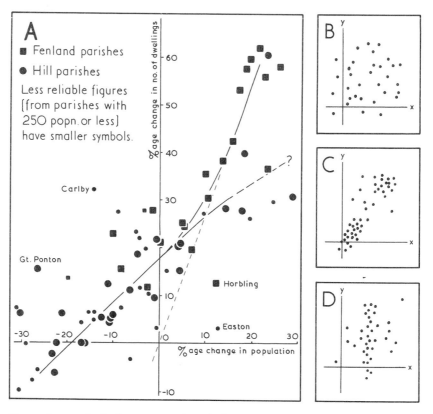

Figure 8   The scatter graph. **A**: A scatter graph to indicate the relationship between percentage change in population and percentage change in the number of dwellings in parishes in south Lincolnshire 1931–51. **B, C** and **D**: Characteristic types of 'scatter'.

Source: **A**: Registrar-General, *Census of England and Wales, 1951*, County Volume, Lincolnshire, London (H.M.S.O.), 1955.

noticeably fall outside these groups, the very isolation of these values stressing their peculiarity. An example of the use of the scatter graph for grouping data will be described later, in Chapter 7.

A rather more usual tendency than that illustrated in Figure 8C is for the plotted values to lie along a line or, more commonly, within a narrow linear zone, as in Figures 8A and D, and where this occurs it is often possible to express the relationship between the two variables in the form of a mathematical formula. To do so it is necessary to replace the scatter of dots by a single line, either straight or curved, which represents the best average fit through the points on the graph. This is the most important part of the whole process, for the accuracy with which the 'best fit' is obtained seriously affects the nature of any subsequent formula. Not unnaturally, mathematicians have developed rather complicated ways of establishing this best fit and of deducing the significance of the resultant line,[1] but even the person with little mathematical training can make elementary deductions from a reasonably drawn line, provided he is content not to interpret his findings too pedantically or regard them as more than a general guide to a situation. For example, if the plotted points lie in a rather broad linear zone the line of best fit will be a fairly weak one with the resultant relationship rather unreliable, and vice versa.

Let us consider a fairly simple investigation of the implications of Figure 8A which was drawn in an attempt to establish whether any relationship exists, in a rural area in south Lincolnshire, between changes in the population of a parish and the number of dwellings it contained, for the period 1931–51. If such a relationship could be demonstrated (and it seemed reasonable to suppose that it would continue beyond 1951) then the results could be used, for example, by the Local Planning Authority as a guide to the number of houses which would be needed in any area to maintain its population, or the possible consequences of allowing only a limited increase in the number of dwellings in an area.

Inspection of the scatter of points on Figure 8A shows immediately that these run in a fairly narrow zone inclined upwards across the graph from left to right and at least in the left-hand portion of the graph a straight line of best fit can be drawn through the plotted values, paying particular attention to the more reliable ones for the larger villages. This line represents the average tendency of the plotted values, indicates that some relationship does seem to exist between the two quantities and suggests important conclusions such as the fact that until an increase of more than about 18% has occurred in the number of dwellings no increase in population is likely; conversely, if the number of dwellings stays stationary, a decrease in population of about 20% can be expected. The line also releases intermediate values, e.g. an increase in population of 10% is usually associated with an increase of about 25% in dwellings.

So long as the line remains straight it is not too difficult to establish a formula which will describe it. Mathematicians say that all relationships which appear on a graph as a straight line will follow the general formula

$$y = mx + c$$

This is simply a shorthand way of saying 'to find the value of the variable on the vertical

---

[1] For further information on this point the reader should refer to the chapter(s) on 'regression' in any standard textbook of statistics.

axis (*y*-axis) take the value of the variable on the horizontal axis (*x*-axis), multiply it by a factor (*m*) and add to it something else (*c*)'. *m* and *c* are fairly easily determined. *m* is simply the gradient of the line, regarded as *positive* if it slopes *upwards from left to right*, and vice versa; *c* is what might be called the 'zero error'—namely, the value on the *y*-axis when the value on the *x*-axis is 0. Thus in Figure 8A the gradient of the line can be measured in the usual way (e.g. in the 'hill' where the line crosses the two axes it rises about 18 units vertically while travelling 19 horizontally) and is 18/19; similarly *c* is also +18 and the formula would read

$$y = 18/19x + 18$$

The advantage of the formula over the graph lies in the fact that it is much easier to memorise and can be used without the actual graph being present.

Although the left-hand portion of Figure 8A shows fairly steady relationships of the kind just described complications arise in the right-hand portion. Here the 'straight-line tendency' breaks down and the zone tends to fork, one half more or less continuing the original trend, but the other curving upwards much more steeply. When this occurs it is worth while going back to the original idea of using a scatter graph to distinguish between different kinds of things and looking to see if there is any logical reason for the distinction between the two groups. Investigation shows that in this case this might be so, for the members of the upper group are almost entirely Fenland parishes while those in the lower one are hill parishes, although here the number is so small that it is difficult to decide whether they constitute a valid trend at all.

Once curved lines are introduced, mathematical relationships are beyond the ability of the layman to deduce (although one could say that in the Fenland villages the trend changes and closely approximates to another formula ($y = 50/18x + 0$), but the example has been taken far enough to show that the scatter graph has a good deal of information to offer. In this case we have obtained a formula[1] which is reasonably typical of the greater part of the range of values likely to be encountered and have had our attention attracted towards lines of further investigation such as the possibility of a fundamental difference in the dwelling-people relationship in Fenland and hill parishes (e.g. variations in the ratio of private to council houses or tenanted to owner-occupied property) and the reasons for exceptional values, e.g. Great Ponton, Carlby, Horbling and Easton, which stand out from the rest. In the investigation of a relationship of this kind, which cannot be other than relatively imperfect because of widely varying local conditions, quite sensible results can be obtained by the layman without any need to introduce higher mathematics.

B   TECHNIQUES SHOWING PROPORTIONS FORMED BY CONSTITUENT PARTS

The second main purpose to which the simpler statistical techniques are devoted is that of indicating the part which various components play in forming the whole. This is normally a relatively simple problem and as several of the techniques are derived from ideas already encountered they can be fairly briefly described.

[1] With a rather general relationship such as the one analysed here little would be lost if this formula, once obtained, was rewritten as $y = $ (approximately) $x + 18$ since 18/19 can be regarded as near enough to unity.

Compound line and bar graphs

So long as the vertical scale on a simple line or bar graph commences at zero it will be possible to subdivide the area beneath the line or contained within the bar into any number of component parts; the result can be called a compound line or bar graph and examples are illustrated in Figure 9.

The basic principle is a simple one, deceptively simple in practice, for the subdivision should be attempted in moderation and with some forethought if the result is not to be mathematically exact but visually disappointing. It is by no means easy on such compound graphs to follow the trend of increase or decrease in all the components. A glance at Figure 9 will show that the only component whose trends can *certainly* be observed easily is the bottom one, which has a horizontal line at its base; the others must be assessed as

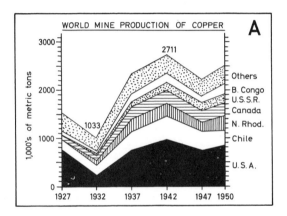
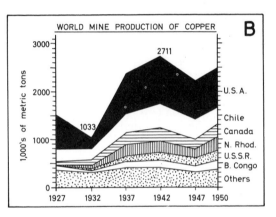

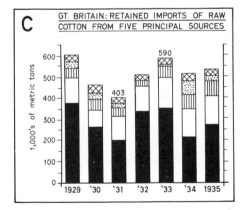
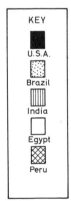
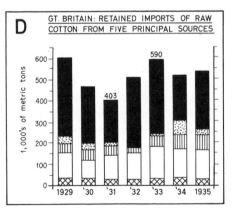

Figure 9 Compound line and bar graphs to illustrate component parts. **B** and **D** portray, in a more satisfactory and legible manner, the information contained in **A** and **C**.

Source: **A** and **B** J. C. Weaver and F. E. Lukermann, *World Resource Statistics*, Minneapolis, 1953. **C** and **D**: L. D. Stamp and S. H. Beaver, *The British Isles* (First and Third Edns.), London, 1933 and 1941.

spaces between pairs of lines, which may be increasing or decreasing at different rates and moving to different levels on the graph, according to the accident of the total of the components which lie beneath them. In the case of the bar graph all the lines remain horizontal but the individual sections may bob up and down making side-by-side comparisons less simple.

The best defence against this is to devote some attention to the *order* in which the components are taken. The order in which they appear in the published statistics is unlikely to be the best one for this purpose and the most satisfactory arrangement is to plot *the most stable elements at the bottom*, thus ensuring that vertical movements are damped down for as long as possible. Figures 9B and D show redrawn versions of Figures 9A and C where this principle has been observed. Both are considerably easier to read than their predecessors, especially Figure 9D where side-by-side comparisons are now quite easily made. Between 1927 and 1932 fluctuations in world copper production were almost entirely confined to the U.S. component, though in Figure 9A this immediately 'tilts' the representation of all the other producers; similarly, and not unnaturally in the largest producer, production in the U.S.A. tended to fluctuate more than in most other countries and the effect of this at the base of the graph is responsible for most of the 'ups and downs' of the other components. If, however, one component is of much more interest than any of the others it should be placed at the bottom of the graph and the general rule of arrangement modified accordingly.

Divided circles or pie graphs

The divided circle, or pie graph as it is almost always called, is one of the commonest statistical diagrams. The total quantity concerned is represented by a circle which is divided into segments proportional in size to the components; comparisons can be made between fluctuations in these components in two or more examples if a circle, subdivided in this way, is drawn for each of them. If proportion alone is important each of the circles can be of the same size, but it is rather more common to use circles of varying size so that the area of each is proportional to the total quantity which they represent. In this way the two elements of quantity and subdivision are both reflected in the result.

Figure 10 shows in this manner how the four main source areas of overspill population in England and Wales have utilised different kinds of reception areas for this population. Comparison between the part played by any type of reception area in the four cases must be made by comparing the *angle* which that sector of the circle occupies in each case.

Perhaps the most unexpected feature of pie graphs is the quite considerable amount of simple calculations involved if they are used in any quantity. Normally, three distinct stages are involved before plotting can begin; these are:

1 a circle must be drawn proportional in area to the total quantity to be represented
2 each component part must be expressed as a decimal or percentage of the whole
3 the angle which corresponds to this decimal or percentage of 360° (for that is, of course, the total angle available for subdivision at the centre of the circle) must be calculated.

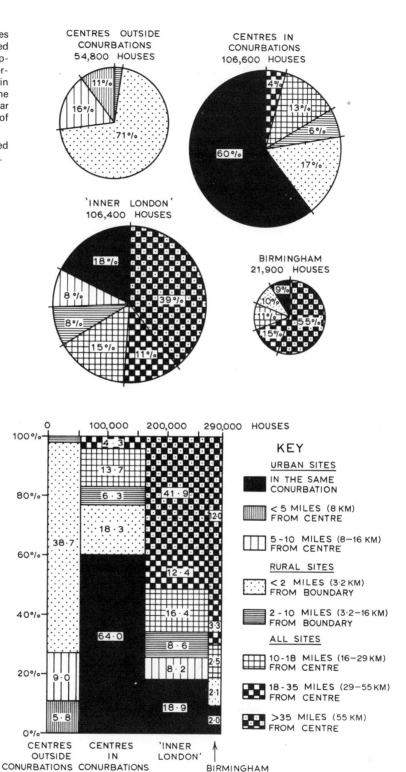

Figure 10 Divided circles (pie graphs) and divided rectangles—types of reception area proposed for overspill from different sources in England and Wales. The figures in the rectangular diagram refer to 1,000's of houses.

Source: Information supplied privately by local authorities.

CENTRES OUTSIDE CONURBATIONS
54,800 HOUSES

11%
16%
71%

CENTRES IN CONURBATIONS
106,600 HOUSES

4%
13%
6%
60%
17%

'INNER LONDON'
106,400 HOUSES

18%
8%
8%
39%
15%
11%

BIRMINGHAM
21,900 HOUSES

9%
10%
1%
55%
15%

0   100,000   200,000   290,000 HOUSES

100%
4·3
13·7
6·3
41·9
80%
12·0
18·3
38·7
60%
12·4
16·4
40%
3·3
64·0
8·6
2·5
8·2
20%
9·0
2·1
18·9
5·8
2·0
0%

CENTRES OUTSIDE CONURBATIONS   CENTRES IN CONURBATIONS   'INNER LONDON'   BIRMINGHAM

KEY

URBAN SITES

IN THE SAME CONURBATION

< 5 MILES (8 KM) FROM CENTRE

5-10 MILES (8–16 KM) FROM CENTRE

RURAL SITES

< 2 MILES (3·2 KM) FROM BOUNDARY

2-10 MILES (3·2–16 KM) FROM BOUNDARY

ALL SITES

10-18 MILES (16–29 KM) FROM CENTRE

18-35 MILES (29–55 KM) FROM CENTRE

>35 MILES (55 KM) FROM CENTRE

Thus 8 entities each with 7 components will mean 8 calculations under 1 (above) and 56 under 2 and 3, but fortunately tables can be used to avoid some of this work.

The simplest of these calculations is that involved in stage 1. To draw a series of circles whose areas are proportional to given quantities all that is necessary is to look up the square root of each quantity and use this number, converted into suitable units such as millimetres or $\frac{1}{100}$ths of an inch, as the radius of the circle. There is no need to introduce π into these calculations since the circles must not have an *absolute* area but only be *proportional* in size. Tables of square roots are normally incorporated in four-figure mathematical tables, usually in the form of two pages, the first giving square roots from 1 to 10, the second from 10 to 100, but in fact these can be used to find the square root of *any* number. Square roots of numbers with an *odd number of digits before the decimal point* should be sought in the pages showing *square roots 1 to 10*, those of numbers with an *even number of digits before the decimal point* on the pages *square roots 10 to 100*. For numbers greater than 100 the printed position of the decimal point in the square root tables should of course be ignored and its true position found by inspection, e.g. to find the square root of 53,426·28.

1   If four-figure tables are being used reduce the numbers to four significant figures, i.e. 53,430·00.
2   There are 5 digits before the decimal point. On the 1 to 10 page the square root corresponding to 5,343 is 2,312.
3   Since 100^2 is 10,000 and $1,000^2$ is 1,000,000 the square root of 53,430 must lie between 100 and 1,000 and the decimal point must therefore be placed to give the answer: 231·2.[1]

Unfortunately, calculation cannot be simplified in stage 2, although a slide-rule will supply sufficiently accurate information quickly and easily here. Stage 3, however, can be greatly simplified either by using a percentage protractor, i.e. one graduated in percentages not in degrees, or by using the table given below. For this purpose the component must be expressed as a decimal of the whole to three significant figures. If these figures are termed, A, B and C the corresponding angle can rapidly be deduced from the table in the manner shown.

Divided rectangles
It is a simple step from the divided circle to the divided rectangle and the basic principle of this idea has already been encountered in the subdivided bar of the compound bar graph. In this case a rectangle (whose area may be made proportional to the total quantity) replaces the circle and can be subdivided into layers, each representing one of the components.

It is rather surprising that the divided circle is much more commonly encountered than the divided rectangle, for the latter has a certain advantage when comparisons have to be made between several examples. This derives from the fact that whereas a circle has only one dimension (the radius) which can be varied, the rectangle has two and variations in area can be accommodated by altering only one of these and keeping the other (usually the height) uniform. If this uniform vertical height is made to equal 100 units of any kind

---

[1] In this case the answer is only an approximate one because of the reduction to four significant figures.

Table for converting decimals of a circle into degrees

A	Table One value used	BC	Table Two value used	BC	value used
1	36°	03	1°	53	19°
2	72°	06	2°	56	20°
3	108°	08	3°	58	21°
4	144°	11	4°	61	22°
5	180°	14	5°	64	23°
6	216°	17	6°	67	24°
7	252°	19	7°	69	25°
8	288°	22	8°	72	26°
9	324°	25	9°	75	27°
		28	10°	78	28°
		31	11°	81	29°
		33	12°	83	30°
		36	13°	86	31°
		39	14°	89	32°
		42	15°	92	33°
		44	16°	94	34°
		47	17°	97	35°
		50	18°	00	00°

example:

Express ·371 of a circle as degrees.
Call ·371 ABC, therefore A = 3 and BC = 71.

If A = 3 value used is 108°.
If BC = 71 no value is given.
Take nearest value to 71, i.e.
BC = 72 when value used is 26°.
Add these two values.
Answer is 134°.

and the components are expressed as percentages, both subdivision and subsequent comparison will be easy; moreover, the several examples can be placed completely side by side within one block, which can also show a scale of percentages on the vertical edge and even, if required, a scale of quantity along the horizontal edge. Figure 10 shows the same information presented in both forms and amply illustrates the superiority of the divided rectangle over the pie graph for side-by-side comparative purposes. It can be argued, however, that if the devices have to be widely scattered, e.g. over the face of a map, comparison is slightly easier between divided circles than divided rectangles.

The triangular graph

The triangular graph is a statistical device which has several facets to its character. In certain respects it can be considered as a variety of graph which is able to show three variables instead of the usual two, and in both interpretation and use it has strong affinities to the scatter graph; alternatively, since the three variables must be expressed as percentages adding up to 100% in total, the graph can be considered as being primarily concerned to show varying proportions as part of the whole.

As Figure 11A shows, the basic idea is once again easily understood. In form the graph consists of an equilateral triangle with sides 100 units long, each carrying a scale running from 0 to 100.[1] Each point of the triangle acts as both *100 for one scale and 0 for another* and the scales usually increase in a clockwise direction along each side. As in the normal graph these scale values are carried across the graph by lines, not in this case all horizontal but inclined to each scale line at an angle of 60°; the direction of the line through 0 on any scale is given by the adjacent side of the triangle and the rest of the lines applying to that scale run parallel to the line through 0.

[1] Special triangular graph paper is usually purchased for drawing these devices, though it is possible to construct the framework oneself.

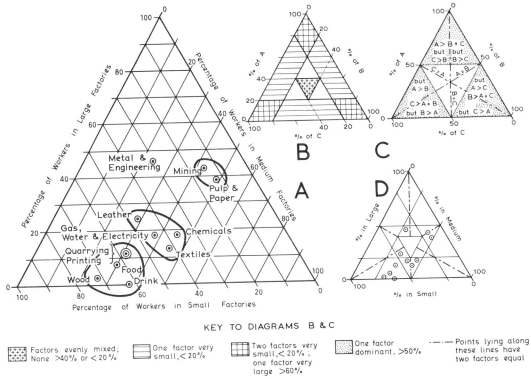

KEY TO DIAGRAMS B & C

Factors evenly mixed; None >40% or < 20% | One factor very small, < 20% | Two factors very small, < 20%; one factor very large >60% | One factor dominant, >50% | Points lying along these lines have two factors equal

Figure 11   The triangular graph. **A**: Proportion of workers in small, medium and large sized factories in the main industrial groups in Sweden. **B** and **C**: Significance of position in different parts of the triangle. **D**: Points plotted on A in relation to areas marked on B and C.

Source: Statistiska Centralbyran, *Statistisk Arsbok for Sverige, 1960*, Stockholm, 1960.

If the scales are correctly drawn and numbered it will be found that any combination of three variables, so long as they add up to 100, will be represented by only one point. In plotting it is best to start with one component, find its value on the appropriate scale and carry this line across the graph until it meets a line corresponding to the value of the next component on its scale. There is no need to bother about the third component since the plotted point will automatically be correctly situated in relation to the third scale.

Triangular graphs are used to study the way in which three components or variables occur throughout a wide series of examples. They can be used, for example, to study variations in the proportion of sand, silt and clay in soils, or primary, secondary and tertiary industry in the industrial structure of a series of towns, provided that these three components form the whole content of the soil or the industrial employment respectively. The most normal requirement when proportions are analysed by means of a triangular graph is to establish the existence or otherwise of groups of similar places so that some sort of classification can be produced. Once these groups are recognised their position on the graph will also suggest a description of their nature, but a little practice is needed before most observers can read the graph easily. Figures 11B, C and D try to give some help in this respect by emphasising the implications of location in certain zones of the graph, e.g. a location close to one of the *points* of the triangle implies that one component must be very large, whilst location close to a *side* of the triangle indicates that one component is quite small. In each case it is wisest to check just which component is large or small carefully from the scales at the side of the graph, but if Figures 11B and C are kept in mind when analysing any scatter of points on the graph they will be found most useful in translating the pattern into everyday language.

As an example of the working of this device Figure 11A can be used to study the way in which employment in the twelve main industrial activities in Sweden is divided amongst small, medium-sized and large[1] factories, and Figure 11D shows the same values against the background of lines which have been used in Figures 11B and C. It will be seen at once that in this respect Swedish industry tends to fall into three groups, with the metal and engineering industry standing rather distinctively outside these. One of these groups, containing the quarrying, printing, wood-working, food and drink industries is characteristically dominated by small factories which never provide less than about 55% of the total employment, whilst large factories are almost entirely absent. The second group, which contains the leather, gas, water and electricity, chemicals and textile industries is not dominated by any single component, but tends to have a fairly low proportion of large factories and a very even division of the bulk of the jobs between small- and medium-sized works. In mining and pulp and paper there is again fairly even division, this time between large- and medium-sized factories with very few small ones.

2.2 Statistical maps

Within the limits of the basic principles of its design, almost every type of statistical diagram allows its designer relatively wide freedom of action. In marked contrast to this, as soon as a map is used as the basis for any composition, a rigid framework is introduced which will exert considerable influence over the design and content of the finished work. For example, the space available for display of any detailed information now depends not only on the overall size of the whole map but on the area on that map within which the information could be placed and still be reasonably representative of its true position. With statistical maps this important element of position forces one to work within rather finer limits than with statistical diagrams, but fortunately a very wide range of techniques is available to help to overcome this problem. These range from very simple ones to those which are precise almost to the point of pedantry; let us consider a very simple type first.

A   NON-QUANTITATIVE STATISTICAL MAPS

No statistical map could be simpler in its design than the non-quantitative variety. Its ambition seems equally simple—namely, to mark in the places or areas where any

[1] In this example limiting values of 100 and 500 employees were used to define the three categories.

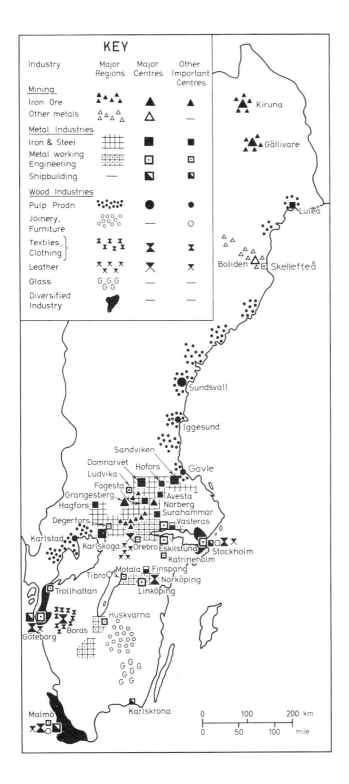

Figure 12 A non-quantitative statistical map—main industrial areas in Sweden (after Bergsten).

features of interest occur, without need to differentiate according to size or importance if any feature occurs more than once.

Because of the very generalised and non-quantitative nature of its information this is a very easy type of map to construct and one frequently met with in atlases and text-books. In its non-quantitative nature, however, also lies its weakness for, as will be seen later in this chapter, the quantitative aspect of various occurrences is often regarded as of sufficient importance to warrant the introduction of many very detailed methods to represent it on statistical maps.

Ironically enough, despite this obvious criticism, non-quantitative maps are often among the most successful statistical devices. Their great merit lies in their ability to *summarise* a situation. Being free from any specific code of representation they can cut away ruthlessly all confusing detail and leave only the fundamental items for the observer to notice.

Much depends here on the care exercised in their preparation. If the information finally presented has been derived by reducing a great deal of material to manageable proportions then they may in fact be among the most discriminating and sophisticated of statistical maps; if, on the other hand, they plot anything and everything they may be almost useless and even downright misleading. Should more information be available than is needed, it may even be possible to overcome the worst aspects of their non-quantitative nature by crudely categorising any items plotted, as, for example, major and minor occurrences, large, medium or small examples, and so on. This has been done in Figure 12 where very detailed information was available from the 'Atlas över Sverige'.

Further situations in which the non-quantitative map has advantages arise where the information to be plotted is in rather variable form (e.g. given in several different units which are not easily interconvertible) of varied reliability or of varied and not strictly comparable nature; e.g. in Figure 12 it would be almost impossible, except in terms of value, to find any way of making quantitative comparisons between, say, shipbuilding and textiles, however detailed a method had been used.

B QUANTITATIVE METHODS

Despite what has just been said in favour of the non-quantitative statistical map, aspects of quantity are often most important in statistical mapping, particularly if the topic under consideration is a relatively simple distribution of only one or two items. The very word 'statistical' implies some assessment of quantity and it is the aim of most published statistics to present this quantitative information to the user in the most convenient manner, and of most statistical cartographers to represent this by means of visually appreciable methods. It cannot be denied that in both cases the results sometimes fail lamentably.

To a certain extent the form in which the actual statistical information is given may well influence the techniques which can be used and the design of the resultant map, for there are three basic types of quantitative statistical distributions with rather different techniques to suit each. These three main types, as indicated in Figure 2, are those in which quantities occur (1) at a series of points, (2) contained within given areas and (3) along a series of lines. Let us consider them in that order.

C   QUANTITIES DISTRIBUTED AT A SERIES OF POINTS

Of all three basic types of distribution this one probably causes least complications; there is, for example, no possible doubt as to the location on the map at which a quantitative symbol should be placed. Unfortunately, statistics relating to this kind of distribution are not so common as is normally imagined. The great majority of published statistics relate in fact to *areas* and many which at first sight may seem not to do so, e.g. information about a particular town, will usually be found to refer to an administrative area centred on that town. Fortunately, the scale of the base map can help in some of these cases and a unit which would have to be treated as an area at a scale of 1 : 50,000 can often be regarded as a point on a scale of, say, 1 : 500,000.

Minor difficulties such as this excepted, the only important problem is to devise a suitable form of proportional symbol which can successfully represent the quantities concerned where they occur. If possible the symbol should be bold, compact and visually appreciable and a variety of solutions have been suggested.

Repeated symbols

One of the simplest of these is the repeated symbol. This can be either of the simple or geometric kind, as in Figure 13A, or of a pictorial or descriptive nature. Thus a given unit of population or labour force might be represented by a diagrammatically drawn human figure, so many million gallons of oil production by a drawing of a derrick or barrel, and so on, with quantities less than the unit chosen being shown by only partially completed symbols.

The great appeal of the method lies in its simplicity—quantities can easily be deduced by counting symbols and the representative nature of the pictorial device makes its message easily understood, even by quite young children. Even so, the method has its difficulties. The pictorial type of symbol may be cumbersome and certainly is often tedious to draw, although geometric ones as in Figure 13A are simpler and fit into smaller space; rather more troublesome is the nature of the symbol, which is in effect a linear one made up of several units. This makes large quantities difficult to handle and counting the symbols only remains practicable as long as their number is fairly small. A solution to this last difficulty is to introduce a second size of symbol representing a larger quantity, so that the number of units involved becomes much reduced; values represented by this means have the great advantage that they can be 'read back' from the map or diagram quickly and with a fair degree of accuracy.

Though the dot map (the counterpart of this method for use with statistics relating to areas) is widely used, repeated symbols *for point values* are not particularly common, perhaps because of the considerable amount of space which is occupied by a compound symbol of this kind with spaces between each unit. A symbol composed of a single unit is inevitably more compact and easier to draw and it is devices of this kind, rather than repeated symbols, which are often used for quantitative maps of point values. Space is very often at a premium on a complicated statistical map and, furthermore, where congestion occurs, symbols 'in bits and pieces' are apt to become confused with their neighbours.

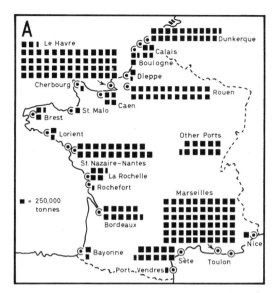

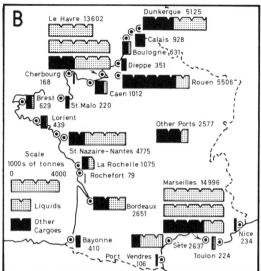

Figure 13 Goods unloaded at French ports in 1958, shown by—**A**: Repeated unit symbols. **B**: Proportional bars.

Source: Institut National de la Statistique et des Études Économique, *Annuaire Statistique de la France, 1959,* Paris, 1959.

Proportional bars

A compact analogy of the repeated symbols used in Figure 13A is the proportional bar of the kind shown in Figure 13B. We have already met the bar in some of its varied forms twice before and the idea here is still basically the same. The bars can be of any convenient width, i.e. broad enough to give a suitably solid symbol but not so broad that they get into one another's way, and can be placed either vertically or horizontally with one end close to the point where the quantity is located.

The bars are simple to draw, flexible to arrange in congested areas and, because of the simple linear form, easy to estimate visually, either unaided or with the help of some sort of scale which can easily be added, as in Figure 13B. Even so the device is not particularly successful and is in consequence not widely used. One great defect is that the linear nature of the symbol makes it very difficult to accommodate a range of quantities of any great extent. A quantity 100 times larger than another needs a bar 100 times as long and this leads either to bars which are so small that they are almost invisible or so long that they have to be broken up into sections which can be 'stacked', e.g. the bars for Le Havre and Marseilles on Figure 13B. Rather more pertinent an objection, however, is the fact that the bar has little sense of place; particularly as its size increases the visual 'weight' of the bar becomes more and more detached from the actual locality it is supposed to symbolise and the message of the map becomes rather vague from a distributional point of view.

The proportional bar is, perhaps, most successful where one end relates to some obvious feature, such as a coastline, so that it is easy to see which end matters and which locality is indicated. Certainly the device is more widely used for representing port traffic than almost any other kind of activity and where the bars can be drawn in extensive areas of sea the extreme length is not so marked an inconvenience.

Proportional circles

The difficulties of the linear scale of values associated with the proportional bar can easily be overcome by introducing a second dimension and producing symbols whose *area* will be proportional to the quantity represented. Circles and squares are the obvious choice and, of the two, circles, which are much easier to draw, are far more frequently encountered. The device has already been met with in principle in discussing pie graphs, and instructions for drawing a series of circles with areas proportional to a given range of numbers will be found under that heading.

The advantages of introducing area instead of length are immediately apparent. Since the area of the figures is proportional to the *square* of the radius or length of side, a symbol 100 times the amount of another need now have a square or circle only 10 times as a large and a much greater range of values can be represented by legible symbols. On the other hand the bulk of the symbols may be troublesome. Even in congested localities it is not difficult to make room for another bar, but a proportional square or circle is not so easily fitted in and overlapping is frequently the only possible solution. Figure 14A, which shows daily passenger traffic at stations in West Middlesex, illustrates a series of statistics where both the values to be represented and the locations of those values almost preclude the possibility of avoiding overlap. Overlap is not necessarily to be deplored, however, for congestion immediately attracts the eye and may be very useful in emphasising important areas in the distribution.

It is usual to add to maps of this kind a scale, either of diameters or of examples of typical values.

Proportional spheres and cubes

The possibility of dealing more easily with extended ranges of values does not end with the introduction of a second dimension. A third dimension exists and, not surprisingly, cartographers have thought of using it to produce symbols whose *volume* would be proportional to the quantity they represent. The result is the proportional spheres or, much more rarely, cubes which are occasionally to be seen in statistical mapping.

Unlike the square or circle the sphere and cube bring rather more disadvantages than advantages in their train. Their great merit lies in the fact that within a given range of symbol size a very much wider range of values can be accommodated. The volume of these symbols is proportional to the *cube* of their radius or length of side and a symbol 10 times larger than another will now represent a value 10^3, i.e. 1,000 times greater, but here the advantages end. Mathematically correct though this relationship may be, the ability of most people to translate its visual representation is very limited indeed and the implications that a symbol 10 times bigger than another represents a 1,000-fold increase are frequently not realised, even with a scale of values to help. This is another example of sticking too rigidly to mathematics and forgetting the limitations of the observer; estimation from bars is easy, from squares and circles more difficult, but from spheres and cubes almost impossible so far as the ordinary person is concerned—a point which will be elaborated further in Chapter 5.

Other difficulties with this method arise from the actual drawing of the symbols, which is far from easy. Cubes, though rarely used, can be represented without too much difficulty if they are drawn isometrically; spheres are more difficult and are usually drawn in one of two conventions, either as 'globes' with a crude graticule of meridians and parallels or as 'illuminated billiard balls' with a highlight near the top of the ball. The spheres in Figure 14C are drawn in this second manner and in both styles the detail is used to secure a convincing three-dimensional effect. Successful achievement of this by either method is tedious and sometimes produces the disconcerting effect of causing the symbol to appear 'floating' above the map and detached from its locality, whilst unsuccessful results look flat and destroy the visual theory behind the method.

Figure 14C shows an example of the use of proportional spheres to plot the same data as Figure 14A. As so often happens the location of the individual values means that it is still impossible to avoid overlap, even with the possibility of dealing with larger numbers more compactly, and the labour involved in drawing the map was very considerable. On the whole the problems of extended ranges are perhaps best tackled by other means than the introduction of complicated mathematical relationships. Chapter 5 gives some possible lines of approach, but other techniques too can be called in to help.

A range of graduated symbols
One such technique is the use of a range of graduated symbols, each symbol representing a specific group of values, the symbols increasing convincingly in size as the quantities they represent get larger. If simple circular symbols are used it is possible to maintain some sort of mathematical relationship by making them proportional in size to the mean or median value of the quantities in the group, thus producing a 'proportional circles map' with a much reduced variety of symbols, but it is seriously open to question whether this preoccupation with mathematical accuracy is really worth all the trouble it is likely to cause. As Chapter 5 will show, it has only limited success visually, and a result no more misleading, much more positive and much easier to draw can be produced by using a well-devised range of graduated symbols (of circular or any other shape) as has been done in Figure 14B.

In West Middlesex the number of passengers using the stations each day varied from 40 at Colnbrook to 31,400 at Golders Green, but all values within this range were represented by a series of eight graduated symbols. The size of these symbols is, of course, quite arbitrary, but was conditioned by such obvious factors as the need for the largest symbols, which represented very large numbers indeed, to be dominant visually, whether this caused overlapping or not; even if strict mathematical relationships are not maintained it would be equally foolish to ignore them completely.

It is also important, since everything has been grouped into categories, that the symbols used for each should be distinct and recognisable on sight. This is very frequently ignored on maps using this device so that it becomes impossible to distinguish quickly one kind of symbol from another, e.g. where a range of, say, ten graduated circles, each only slightly

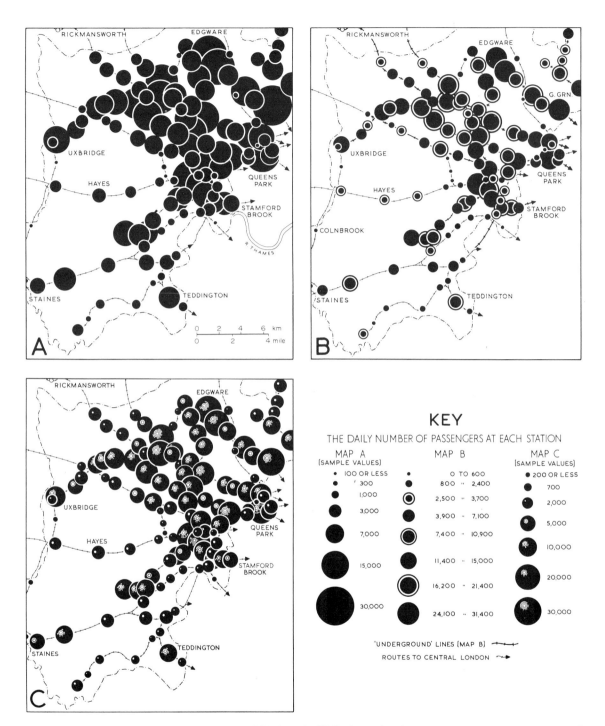

Figure 14 Passenger traffic at stations in west Middlesex in 1951, shown by: **A:** Proportional Circles. **B:** A Range of Graduated Circular Symbols. The division points between the symbol groups have been established using a dispersal graph (see Chapter 4, p. 87). **C:** Proportional Spheres.

Source: County of Middlesex *Development Plan, 1951, Report of the Survey.*

larger than the next, is employed. Figure 14B attempts to avoid this by achieving graduation through an intermediate type of symbol, the 3rd, 5th and 7th symbols in the range being made by adding to the preceding symbol an outer circle equal in diameter to the succeeding one, thereby achieving a transition which is gradual and visually distinct. This is only one way of making this distinction, however, and many other devices such as a notched or marked symbol could be used.

The gradual range of symbols has several easily recognised merits. It is easy to draw, avoids calculations and allows much wider freedom in accommodating the range of values; equally as important is the fact that, where a large number of quantities are represented, as in Figure 14B, it makes appreciation of the total message of the map easier by sorting out for the observer one of the essential aspects of the distribution. Whereas with an infinite number of symbols the *difference* between quantities is always being stressed visually, *whether it matters or not*, a limited range of symbols stresses *the equally important similarities* which many of these quantities may share. Thus in Figure 14A Hayes stands out as rather different from its neighbours, whereas Figure 14B shows that, *in terms of the kinds of variations encountered in the area*, this difference is relatively unimportant.

This 'sorting-out' process, which we have encountered here for the first time, is extremely important in many statistical techniques and, well done, usually makes a significant contribution to good map design. If we are to take many individual values and arrange them into groups, emphasising broad similarities and ignoring minor differences, it goes without saying that the values which we choose as dividing points between the groups are most important and will play a critical part in the appearance of the resultant map. Unfortunately many amateur cartographers (and, to judge by the maps one sees in books and atlases, many not-so-amateur cartographers as well) are not convinced of this and feel that all that is needed is to slice up the range of values into groups by arbitrarily choosing convenient round numbers as division points. It cannot be stressed too strongly that *this simply is not good enough*. The division process must be done with care and there are several methods available for doing it. So important is this process in fact that it has been singled out for separate treatment and will be dealt with in Chapter 4; no more will be said about it here but it must be remembered that a satisfactory understanding of the nature and characteristics of the 'range of graduated symbols' technique, and of other techniques such as shading and isolines which follow later in this chapter, will be impossible unless the relevant portion of Chapter 4 is also consulted.

D QUANTITIES CONTAINED WITHIN GIVEN AREAS

Statistics showing quantities contained within given areas are perhaps more common than any other type, but unfortunately their use is attended by several difficulties of a general nature. Chapter 3 is concerned with elaborating some of the finer aspects of these difficulties, but their broader aspects can well be introduced here. Foremost amongst them is the knowledge that the statistics simply indicate that within a given boundary-line a certain number of items are to be found; just how these objects are arranged within that boundary-line is not indicated—the distribution may be even or alternatively extremely uneven, and if the area concerned happens to be a large one such as a national territory then variations of this kind may be very marked indeed. Further information may bring to light more detailed distributional knowledge, but it is often the case that the only secondary sources available, e.g. maps of various kinds, are nonquantitative and leave the cartographer the problem of either making his own subjective estimate or working with average values for the whole unit, though he may know these to be unrepresentative or misleading.

Generally speaking the smaller the units used the less misleading are the figures likely to be, but it may be difficult or impossible to obtain a map showing the boundaries of the smallest administrative units for which figures are given. Ward boundaries in urban areas of Britain are a good example of this elusive type of information and if one is working on a period, say a century ago, with several boundary changes intervening it may prove surprisingly difficult to establish the boundaries of much larger units than wards; published statistics rarely carry a map showing the boundaries of *all* the areas for which information is given and naturally the problem is much worse if the information is required for units outside one's own country.

Very frequently it is impossible to obtain entirely satisfactory solutions to general problems such as these which arise from statistics relating to areas, but the results need not be too disappointing. Provided always that one approaches maps of this kind with a reasonable awareness of the difficulties which these statistics bring and the limitations of the cruder methods relating to larger areas the various techniques described below have a most useful and necessary part to play in statistical representation.

Dot maps

Few ideas could be simpler than the principle underlying the dot map, and yet few statistical techniques raise more ancillary problems before they can be put into practice. In theory the dot map is nothing more than the repeated symbol, which we have already met, applied to areas and confined to only one type of symbol, the dot. If each dot represents so much or so many of a particular item a simple division sum will indicate how many dots will represent the quantity contained within a given area and these dots must then be placed within that area on the map.

This theoretical statement of the method is deceptively simple, for it has already introduced three unknowns, each of which can be varied and needs to be solved before mapping can begin; these are:

1   How much or how many shall each dot represent?
2   How big shall each dot be?
3   Exactly how shall each dot be placed within the area?

A fourth problem, how shall each dot be drawn, could also be added.

The answer to the first of these problems must be suggested by the statistics themselves. Generally speaking, it is desirable to have at least two or three dots for each area so that within each area there is at least a crude pattern of dot distribution; it is with variations in this pattern from one area to the next that the message of the map begins. The great difficulty lies in finding a dot value which achieves a satisfactory compromise by producing this effect at the lower end of the range and yet does not demand an excessive number of dots for the larger quantities which will almost inevitably occur.

The second question is even more subtle and will to a great extent depend upon decisions just described and on methods available for drawing the dots. It should be remembered that, within the dot map, variations in the density of dots over the face of the map must be judged by changes in the ratio of solid dots to white background, and if this is to be successful as wide a variation as possible should be introduced into the amount of the 'black' dot component. A useful working rule is to try to arrange the number and size of the dots so that they will just begin to coalesce in the densest areas; this implies that since any area where no dots are needed will be 'white' the map will carry all variations from white to almost solid colour and the eye will have this wide variation to help in estimation. The point is well illustrated in Figures 15A and B, where exactly the same distribution has been plotted twice, once with large dots and again with smaller dots. It is immediately apparent that in Figure 15B it is much more difficult to judge variations in density, and the smaller amount of black on the map tends to reduce everything to a more even-looking distribution. In Figure 15A the congestion produced by the bigger dot is a decided advantage, emphasising and attracting attention towards the areas of densest distribution.

One of the most difficult aspects of using larger dots lies in finding some satisfactory way of drawing them. The best method is undoubtedly to obtain a special pen nib which has the end rounded and bent at an angle to the rest of the nib so that each application produces one dot (see Figure 43, p. 128 for an illustration of such a pen). These are known as *dotting-pens* and come in a wide range of sizes; similar to these are pen nibs supplied for lettering but which have the rounded part bent at right angles to the rest of the nib. These also produce a dot if merely applied to the paper without movement. The dots in Figure 15A were drawn with one of the larger pens supplied for use with 'UNO' stencils. These again form a dot if merely touched to the paper, but the larger pens are very free-flowing and will 'blob' unless only a very small quantity of ink is placed in the reservoir; if this does not cure the trouble, it will help to touch the point on blotting paper before each application.

Other methods which can be used if these technical aids are not available are stick-printing of dots, sticking punched dots on to the map or punching out holes in the map which is then placed over a backing sheet. The best and fullest account of the technicalities of these other methods will be found in Birch, *Maps*,[1] but they will not be elaborated further here. The layman will probably find one or another of the three types of pen described above more easily obtainable and quicker in action, while the experienced draughtsman will already have established his preferences. The use of a springbow compass for drawing *small* dots is not recommended to the inexperienced draughtsman, for it will be found difficult to produce consistent, circular results. If crayon or pencil techniques are being used a useful, easy and quick way of drawing large dots will be found by taking a piece of thin card, punching out several holes with an ordinary filing punch and using these as a stencil by first 'drawing round' the hole and then filling in whilst the card is still in position. Holes wear relatively quickly, but new ones can easily be added to replace them.

[1] T. W. Birch, *Maps—Topographical and Statistical* (Chapter 22), Oxford, 1949.

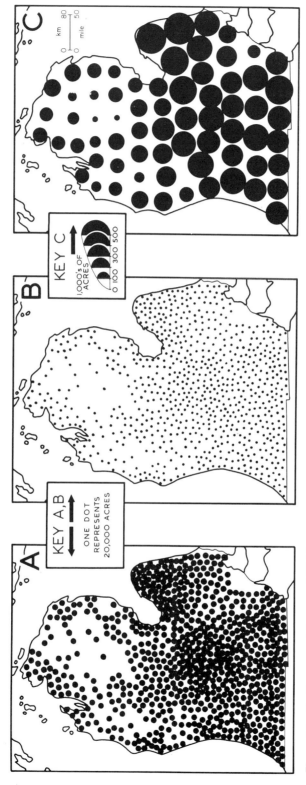

Figure 15 Southern part of the state of Michigan, area of land in farms, 1954, shown by: **A** and **B** Dots, working from a basis of constituent counties. **C** Proportional Circles for each county. In **A** and **B** the information portrayed is identical except for the size of the dot used, illustrating the marked advantage of the larger dot in assessing variations in the distribution pattern.

Source: U.S. Bureau of the Census, *1954 Census of Agriculture,* Washington, D.C., 1956.

In at least one type of map the problem of dot size can sometimes be eliminated. If the dot represents an *areal* quantity then it can of course be made true to scale and the map will give a realistic impression of the amount of land actually occupied by the feature under discussion. The calculations involved are not difficult. Determine beforehand (*a*) the scale of the base map, expressed as so many km to 1 cm—call this quantity S and (*b*) the value in hectares which one dot shall represent—call this quantity D. Then to find the radius (R) in centimetres of such a dot on a map of this scale a formula can be derived as follows:

1   The scale is S kms to 1 cm, therefore 1 square cm will represent S^2 square kms or $100S^2$ hectares (there are 100 hectares in 1 square km).

2   By proportion if $100S^2$ hectares occupy 1 square cm on the map D hectares, the dot value, will occupy $\dfrac{D}{100S^2}$ square cm. This is the area of the dot.

3    $\pi R^2 = \dfrac{D}{100S^2}$ therefore $R = \sqrt{\dfrac{D}{100\pi S^2}}$

$\sqrt{\dfrac{D}{100\pi S^2}}$ can be rearranged as $\dfrac{\sqrt{D}}{S} \times \dfrac{1}{\sqrt{100\pi}}$ and since $\dfrac{1}{\sqrt{100\pi}} = \cdot 0564$

$$R = \frac{\cdot 0564\sqrt{D}}{S} \text{ cms or (better) } R = \frac{0\cdot 564\sqrt{D}}{S} \text{ mms}$$

In imperial measure if S is in miles to an inch and D is in acres

$$R \text{ (in inches)} = \frac{\cdot 0223\sqrt{D}}{S}$$

Once the value of R is calculated it would be reasonable to use any dot diameter which approximated to this quantity without reproducing it with mathematical accuracy.

It does not necessarily follow that where the dot symbol is used to represent areal quantities this procedure need always be followed. A great deal depends on just how much of the total surface the use under discussion actually occupies. In Figure 15A the land in farms not surprisingly totals over 80% of the area of many countries in Michigan and therefore true areal representation will produce the desirable coalescence of dots which was mentioned earlier. On the other hand, had the map tried to show the distribution in the same state of, for example, orchard land, which will not occupy such a large area even in the most important fruit-growing districts, the result would have been too 'thin' and a dot 'larger than life' would have given a better result.

Whatever dot value and dot size are used it should be borne in mind that an area full of dots on the map will carry a psychological impression that on the ground the same area is full of whatever is being mapped, and in the same way too sparse a sprinkling of dots conveys an unmistakable impression of emptiness. Aspects such as this, which arise from decisions on dot size and value, may prompt some modifications to the general rule of trying to achieve coalescence in the denser areas if this would be quite out of keeping with reality.

The problem of placing the dots within the area is, if anything, more difficult to solve than those arising from the two variables so far discussed. Two fundamentally opposed solutions are possible within any given area: either (a) spread the dots evenly, or (b), if variations are known to occur within the area, attempt an uneven distribution of the dots so that they are as representative as possible of these variations. The latter is easily the best solution. The results will be more realistic and will exploit the one very important asset of the dot symbol—namely, its flexibility. Unfortunately, distribution of dots within areas on this basis will almost certainly call for a good deal of guesswork and subjective judgement and the results will vary in reliability depending on the cartographer's detailed knowledge of the topic which is being mapped and the number and nature of any sources which he may have to help him in his judgement. Help of one sort or another will frequently be available for this problem, but unfortunately it will often be qualitative rather than quantitative, so that whilst it may, for example, indicate firmly where *not* to place the dots it will give no strong positive guide as to numbers in those areas where dots are needed. Maps are among the commonest and most easily available aids of this kind: e.g. even a simple topographical map at a scale of 1:50,000 or 1 in to 1 mile (1·6 cm to 1 km) would assist in placing dots representing a population distribution by indicating where villages and hamlets were but not, of course, the detailed population of each unit; similarly it might help in plotting agricultural statistics by marking areas of marsh, moor, forest, towns and industry which would need to be left blank on the map. More specialised maps may be even more useful.

Subjective placing of dots in this way has been criticised as introducing into the dot map personal judgements which cannot easily be checked, and whilst this cannot be denied it seems reasonable to point out that the errors involved need be no more misleading than those which are visually unobtrusive but so well known to be present when the dots are spread evenly; moreover, a person constructing a dot map must surely have some additional knowledge of its subject matter beyond the bare statistics and his errors will at least be associated with the intention of taking a step in the direction of establishing the true distribution.

Spreading the dots evenly throughout an area overcomes these difficulties, but often produces a misleading 'average' picture. Used in this manner dots become a kind of infinitely variable shading indicating density within the area but with the major disadvantage, in comparison with other methods which will be described shortly, that they can give no direct indication of this density themselves nor is it easy to add to the map a suitably graduated scale against which comparisons can be made. It is seriously open to consideration whether shading or proportional shading of areas would not be more effective a method than dots spread evenly. Furthermore, it is not easy to achieve an even distribution of a given number of dots within an area and it is always advisable to plot them in pencil first and check for a satisfactory result before inking in. If small numbers of dots are required their locations are easily 'planned' in advance, but if larger numbers are needed the best method seems to be to begin by making an open, even distribution of a few dots and then gradually filling in the spaces, working continually over the whole area all the time. Starting at one end and hoping that the correct distribution has been hit upon so that one will just run out of dots by the time the other end is reached will rarely give satisfactory results.

It is a bad policy, too, to fight shy of placing dots near the boundary line of an area. The centre of the dot is its true location and dots can and should overlap boundary lines if this is necessary. Very often these boundaries are erased from the finished map and unless this precaution is observed the former position of boundary lines will often stand out as empty lines on the map.

In spite of all these difficulties the dot map has been widely used. Its basic idea is attractively simple, no other method relating to areas has the same flexibility and it has a particular appeal from the discontinuous nature of the dot symbols which portray satisfactorily many distributions which are themselves discontinuous in reality, e.g. land under various crops. On the other hand, of all the methods described in this book it is the least positive in its quantitative message, for though the individual dot is quantitative enough one soon wishes to make comparisons of quantities in areas which contain tens or hundreds of dots and this is quite impracticable. The dot map, in effect, shows variations in distribution admirably, but portrays quantities very weakly, so that it is best used in showing distributions which are patchy and where marked variations occur. Thus Figure 15A is able to show well both the patchy distribution of land in farms in the northern half of the peninsula and also the contrast this provides with the southern portion of the state. If the distribution is a fairly even one with variations which are *relatively small but nevertheless important* the dot map will not be a very suitable method to use. The apparently even distribution of land in farms which Figure 15B shows in the south-west corner of Michigan, for example, can be contrasted with the manner in which certain differences are clearly picked out where the same information is shown by shading on Figure 16B (p. 52).

It has also been demonstrated that with dot maps impressions of density of dots in any small area on the map may be markedly influenced by the number and arrangement of the dots which surround that area, but this seems rather an academic point in view of the general weaknesses in legibility in this and many other methods, a subject which will be further described in Chapter 5.

Shading maps (choropleth maps) [1]
Shading techniques present, on the whole, simpler and cruder methods of portraying statistics relating to areas than do dot maps. Both of these characteristics derive from the fact that shading methods *presuppose uniform distribution of the quantity throughout the area*, aspects of quantity and area, for example, being combined by calculating the *average density* of the distribution within the area and representing this density on the map by some form of shading applied to the whole area.

It cannot be stressed too heavily that with shading techniques *the effect on quantities caused by the varying sizes of the areas used must not be allowed to intrude*. The most usual way of overcoming this difficulty is to express quantities as a *density* per square kilometre, acre,

---

[1] Shading maps are often called *choropleth* maps. It is unfortunate that the essentially simple techniques of statistical cartography have been given a variety of complicated, confusing and not-yet-standardised names of pseudo-Greek origin. With the exception of the term *isoline* (see later) these terms are deliberately omitted from this book. It is the author's opinion that they suggest a *mystique* where none exists and tend to induce a rigid rather than a flexible approach to the various techniques.

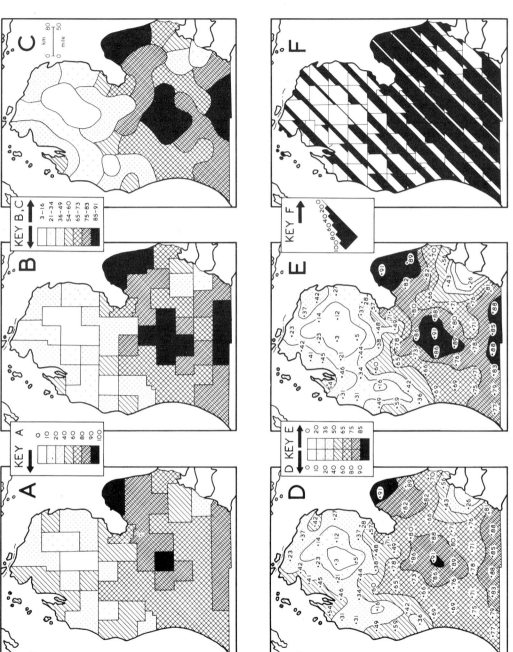

Figure 16 Southern part of the state of Michigan, percentage of land in farms, 1954, shown by: **A** and **B**. Normal shading techniques. **C**. Normal shading techniques modified by 'rounding off' boundaries of constituent areas. **D** and **E**. Isolines. **F** Proportional shading. Since all six sections derive from the same basic information this figure gives an opportunity to compare the effect of using different methods for the presentation of a single piece of material. The effect of variations *within* each method can be seen by comparing **A** with **B** and **D** with **E** ; these show the different impressions of distribution which can be produced simply by varying the limiting values of the different types of shading which are employed.

Source: U.S. Bureau of the Census, *1954 Census of Agriculture*, Washington, D.C., 1956.

etc., or as some quantity which is entirely independent of area such as a *percentage, ratio or per capita figure. Straightforward numerical quantities should never be shown on shading maps.* The reasons for this are almost self-evident, but the author has found that students are persistently tempted into trying to show quantities in this way, frequently in an attempt to avoid the calculations necessary to turn quantities into densities. Imagine a perfectly uniform distribution in which are found two adjacent units, one twice as large in area as the other. The former will obviously contain twice as much of the feature being studied as will its smaller neighbour, yet if shading by *quantity* were applied, these two units would be sharply differentiated; in fact the distribution is uniform, one only containing a greater quantity by virtue of its greater size. A map made in this way would reflect variations in the area of the units used and not, as was its intention, variations in the distribution of some particular feature.

Having expressed quantities for each area in some suitable form, some means must then be found for showing these on the map. The most widely used method is to divide the range of values into suitable groups and distinguish all values falling within that group by a particular type of shading. As with the range of graduated symbols described earlier this division into groups is important. The selection of the necessary dividing points between groups will affect the appearance of the resultant map very considerably (compare for example Figures 16A and B) and should be made with careful consideration using one of the various possible methods described in the first part of Chapter 4 (pp. 82–91). This is no trivial decision. On a shading map the areas coloured-in in any particular way are automatically given a certain 'unity'; their common shading tint says, in effect, 'we are a group of areas who have certain numerical characteristics in common' and, equally importantly by implication 'we are also rather different from all other groups of values.' It is important that this message should have its roots in numerical reality (see Chapter 4) and not merely in constructional convenience.

The divisions established, a suitable range of shading must be devised. An infinite number of variations is possible here, but certain general rules can be applied. The most important is that the shading, in its increasing density of tone, should have a visual resemblance to the kind of numerical increase which it represents. An ideal solution would be to control the density of the shading (i.e. the black-white ratio) mathematically, so that shading representing densities twice as great as those elsewhere would be twice as dark in tone. This could easily be done mechanically with prepared dots and stipples and also by various hand methods which will be described later, but the fact that one type of shading usually represents a group of values, which would have to be replaced by their mean or median, weakens the apparent exactitude of the method, and experiments have shown that, unfortunately, a range of shading which increases evenly in density from white to black does not in fact *appear* to do so to the eye.

In practice there is no need to stick slavishly to mathematics; a reasonably devised increase in intensity will serve most purposes and it is much more important that each type of shading should be distinctive enough to be recognisable on sight. This places a limit on the number of divisions that can be used, although much will depend on whether colour or black and white only is available. Whichever is the case, a maximum of eight to ten types seems suitable in practice. More variations than this, even where each is suitably distinct, gives the eye rather too much detail to look at although a greater number might be justified if it were intended that the map should be used as a statistical 'quarry' rather than as a picture of a situation.

As with dots, the overall impression of variations in shading is important. A range of shading which has very little gradation in intensity will produce a psychological impression that the values it represents show a similarly small range of variations, even though several subdivisions may be distinguished; a wide range of shading tone gives the converse effect and any such psychological impression should not be at variance with the figures.

It is not essential that the two extremes of any range of shading should be 'white' and solid colour. Quite apart from the psychological implications of these two particular tones—the one implying emptiness, the other extreme plenty—they are particularly strong visually, being more easily recognisable than the other types of shading and tending to dominate the map. This may or may not be a satisfactory state of affairs. Frequently it is important that extreme values *should* be emphasised, but once again the shading should be matched up with the figures and the impression it is required to produce.

If colours are available it is usually advisable to limit the number to two or three and build up variety by using several tones in each colour rather than several colours. Once again the balance is a delicate one. It is not at all easy visually to identify at first glance any one of say six shades of yellow, and in this case it would be better to use three yellows and pass on into three shades of brown or orange; the transition from yellow to orange or brown will need making with care if a gradual series is being attempted, but could be kept quite striking if this point in the series marked a major break. Where several colours are used it is usual to pass from one to another in spectrum order, i.e. red, orange, yellow, green, blue, indigo and violet, though it will be found better to keep the latter half of this sequence for negative values. Between yellow and green there is often a more marked colour break than in the rest of the sequence and the arrangement accords well with the established psychological properties of colour, red being a dominant colour, violet a recessive one.

In practice it is naturally easier to obtain suitable tones by using colours which can be mixed with one another to produce transitions. Water colours and poster colours come off best here whilst media which have 'fixed' colours, such as crayons or bottles of coloured inks used without mixing, make it difficult to achieve the required gradation of tone.

If black and white alone are used shading can either be applied by hand or by the use of ready-printed thin adhesive cellulose sheets which can be cut to the required shape and applied to the map. This latter method is becoming very widely used and a surprising variety of shading, stipples and other patterns and devices are available.[1] Unfortunately for the layman it is rather expensive, tedious if the areas are many and small or with irregular boundaries and the sheets are not widely stocked, except in large cities. The result, however, is very neat and 'professional' in appearance. As an alternative, quite satisfactory results can be obtained by hand, building up patterns from lines and dots. As with dot maps it should be remembered that it is the ratio of black to white which gives the impression of intensity and this should be varied as much as possible, e.g. by *thickening*

---

[1] For example those sold under the trade names of *Zip-a-tone* and *Letraset*.

lines as well as by adding more to the pattern. To achieve an even result it is best to place the map on a tracing table over a sheet of graph paper which will act as a guide for the spacing of the lines or dots; rows of dots are more successfully drawn if made with the pen touching a bevelled straight-edge every time it is applied to the paper.

Several of these points are illustrated in Figures 16A and B which indicate the percentage of land in farms in counties in part of the State of Michigan. The juxtaposition of Figures 16A and B emphasises the rather different impressions which are produced when, still using the same number and types of shading, different values are chosen for the limits of the shading types. In Figure 16A the values used were those taken from the 'Land in Farms' map for the whole of the U.S.A. produced by the Bureau of the Census. Though these may be very suitable for dealing with the whole country, and may have to be used if other states were to be mapped in a comparable way, they are *not* suitable for an isolated map of the distribution in the State of Michigan alone. In Figure 25 (p. 88) where the distribution of values is analysed, it will be seen that 80 and 90, for example, *bisect* natural groups and the dispersal graph suggests the much more logical subdivision used in Figure 16B. Figure 16B also shows that where the map does not form part of a series there is no need for the key to be numbered continuously; it is more indicative of the actual state of affairs if the range of each group is given as lying between existing maximum and minimum values, rather than 'theoretical' ones as in Figure 16A.

The seven shading types used in Figures 16A and B were based on simple, hand-applied line and dot patterns with rather a marked step in visual intensity between the third and fourth. This helps particularly to emphasise the contrasts between the northern and southern parts of the peninsula in Figure 16B, but this effect is almost lost in Figure 16A where the line patterns begin at a lower value.

The shading map has one great disadvantage which we have already touched upon—namely, the presupposition that density over a whole area is uniform. This problem is similar to that which was encountered when the distribution of dots within areas was being discussed and it can be resolved in the same way, by making some sort of subjective assessment of the variations which probably exist. J. K. Wright has suggested such a method where one has statistics for a unit which is known to contain two quite different types of territory.[1] Briefly described, the procedure involves first delineating these subdivisions on the map and then making a reasoned guess at the density which exists in one of them. This can be done particularly by comparison with areas of a similar type which may form complete units and for which typical density values would therefore be known. If such a density is then supposed in one subdivision whose area can be measured, the total quantity contained in that subdivision can be calculated (area × density) and the result subtracted from the total quantity for the whole unit; this will yield the quantity in the other subdivision for which the area and then the density can be calculated. Once again, although the method involves an element of personal judgement, the results, when made by a well-informed person, are not likely to be more misleading than the unadjusted values. As an example of an area where this kind of calculation is justified

[1] J. K. Wright, 'A Method of Mapping Densities of Population with Cape Cod as an Example', *Geographical Review*, vol. 26, New York, 1936.

there can be quoted the Urban District of Pickering in North Yorkshire, which, though essentially centred round a small market town on the edge of the Vale of Pickering, is extremely elongated, stretching up in a tongue nine miles long almost to the crest of the North York Moors, and including in its northern end a considerable area of hamlets, scattered farms and moorland. It would be misleading to colour in over the whole of such an area urban characteristics such as population density which apply to only part of it. Fortunately, adjacent to this northern portion there are several entire rural parishes which occupy similar territory, and whose densities of population would be a useful guide in estimating a figure for this northern area.

A further disadvantage of shading maps is that boundaries assume a degree of significance out of all proportion to their true importance so far as a particular distribution is concerned. Abrupt changes of shading which occur at boundary lines hardly ever have any counterpart in reality, and equally stupidly this marked break has to follow every minor inflection and promontory of the boundary itself; detached portions of units are a particular inconvenience here though they can sometimes be dealt with in the manner described above. In matters such as this, the inflexible shading techniques show up poorly against the discontinuous patterns made by dots, which effect rather less noticeably abrupt transitions, but to some extent this disadvantage can be partially overcome simply by smoothing and rounding off the shapes which would result if the boundaries were exactly reproduced. Figure 16C provides an example of this with the values which were used in Figure 16B treated in this way. In Figures 16A and B the rectangularity of administrative units, though odd to European eyes, produces blocks of shading of reasonable shape so long as units do not occur diagonally when an odd-looking pattern results; it seems reasonable to smooth this off into a diagonal band or, for example, to turn a cross shape into a circular one of rather smaller dimensions.

Proportional shading maps

Shading techniques, like dot maps, are widely used despite their disadvantages. Their particular asset is the positive nature of their message, clearly placing the quantity being represented within a given range of values, and so far this is as much as we have expected of them. It is possible, however, to devise shading techniques which will not merely place a value within a range but actually represent it, 'true to scale' as it were. This possibility has already been touched upon in considering mechanical shading patterns, but no complete range of these is likely to be produced as a commerial proposition. Hand techniques, however, can provide a suitable equivalent, and two ways have been suggested.

The first of these is more akin to mechanical methods, but tedious to draw. It requires bands of a convenient width, say 20 mm or 1 in, to be ruled horizontally across the map, these bands being then filled in with a number of horizontal lines according to a scale of one line for a particular number of items. An example will make the point clear. Let us suppose that a map is being prepared to show the density per square mile of workers in agriculture, and that the values to be plotted range from 200 to 20: a scale of one line for every ten workers might be decided upon. Twenty lines would then have to be inserted between each pair of horizontal lines in the area where the density of 200 occurred but only 2 where the density was 20. This would involve dividing the space between the main horizontal lines into 20 equal parts in one case, 2 in the other, 17 if the density was 170 and so on, each density being brought to the nearest ten before work commenced. The tedium arises from repeating this process of division for say sixty different areas. The actual process of division is not in itself difficult and several methods are available, whilst any densities which are repeated can be easily drawn in by 'transferring' the spacing from the first example.

Although the intensity of the lines is indeed proportional to the quantity represented, it is not easy to make comparisons from one area to another; and in effect the discontinuous horizontal lines which make up this type of shading have become as incomparable one with another as the discontinuous dot pattern on the dot map. A scale showing typical values is usually added beside such a map to help this comparison.

The author has found that a second method, rather easier to draw, is also more successful. As before, this needs bands ruling across the map, but in this case they need not be horizontal and they should be of a width of rather more (usually the next convenient round number) than the largest quantity to be represented; at right angles to these is drawn a line along which a scale of values in between the bands can be measured. As an example, if the largest value to be shown was 231 bands 250 units wide would be a suitable choice; in Figure 16F where the highest value was 92 the bands were 100 units wide, the actual physical width being 20 mm on the original drawing so that 1 mm represented 5 units on the scale line. Any value occurring within a given area can be represented by measuring up from the base of *each* band crossing the area a distance equivalent to that quantity measured on the scale line; a line is then ruled through this point parallel to the base of the band and the area between these two lines shaded in solid colour. A glance at Figure 16F will make this quite clear, e.g. where the density was 86, a distance of 17·2 mm was measured above each base line, producing a black band 17·2 mm wide alternating with a white one 2·8 mm wide across all areas where this value occurs. It should be noticed that not only base lines passing through an area but those passing close to the boundary must be considered as well, for a large quantity based on these may well bring the solid colour within the boundaries of the area although in this case filling-in extends only to the boundary and not to the base line.

The method is not difficult to use, but one or two points need some consideration. Foremost amongst these is the width of the bands which in turn will be affected by the size of the areas for which values are given. It is desirable that the band-width should not be greater than the height of the smallest area, although smaller band-widths involve more work and less easily made measurements along the scale line. A compromise may have to be made and it may be quite satisfactory to arrange the bands so that the smallest areas lie astride two bands where they cannot occupy a whole one; alternatively, tilting the bands may allow them to take advantage of some more favourable dimension in the smaller units. It is often helpful to rule the bands out first on tracing paper which can be juggled about over the map to find the most favourable position.

The result produced by this technique is a map which often has quite wide variation in the black-white ratio and a black component proportional to the quantity represented. Comparisons of value from one unit to another can easily be made, particularly between adjacent areas, by comparing the heights of the black bands in each case; a scale is easily added (see Figure 16F) and although boundary lines are not eliminated, the absence of continuous colour means that their effect is more broken up and less obtrusive. A defect of the method is that the scale of shading is a linear one and, as such, may give difficulty in dealing with extended ranges of values.

Figure 16F shows an example of a map prepared by this method, and it is interesting to contrast the picture it presents with the various impressions deriving from Figures 16A to E. It offers, among other things, the first attempt to show the obvious contrast between the northern and southern halves of the peninsula in its true proportions.

Isoline[1] *maps*

The 'unreal' effect which boundary lines produce on shading maps remains a major disadvantage. Smoothing off these boundaries may help (see Figure 16C), but will not remove the problem of abrupt transition in values at a particular line. The isoline technique offers one way of overcoming both these disadvantages. As with shading techniques, it too starts out with an average density or similarly expressed value for each unit but regards this value as being *typical of* rather than *confined exactly* to the area under consideration. Each quantity is plotted on the map within its area as a 'spot-height' value and in between the 'spot-heights' 'contours' are drawn to show the trends of the figures. Between the contours (isolines) layer colouring or shading can be added for further emphasis. The analogy with layer-coloured relief maps is very strong, but should not be pursued too rigorously.

The method involves the solution of three main problems: (1) placing the spot-heights, (2) deciding on suitable values for the isolines and (3) drawing in the isolines. The first of these is not particularly difficult; an obvious solution is to place the spot-height in the geographical centre of the unit, but if more detailed information of distribution within the area is known it will be found more satisfactory to attempt to place it at the 'centre-of-gravity' of the unit, much nearer to any local areas of particularly dense distribution. The second of these is an old problem in new guise. The values selected for the isolines perform the same function here as those which divide different kinds of shading in a shading map, and they should therefore be chosen with the same degree of care, using one of the methods described in Chapter 4 (pp. 82–91).

It is the last of these three problems which presents the cartographer with the more difficult situations so far as this method is concerned, although the basic principle is quite elementary. Isolines which occur between any two adjacent spot-heights are fixed by drawing a straight line between the two points and spacing the isolines along this line proportionately to the difference between their value and that of the spot-heights. Thus if two isolines for 55 and 68 have to be drawn between a pair of values of 42 and 89 they will cut this line at points 13/47 and 26/47 of the total distance between the points measured

[1] The pseudo-classicists have had a field-day here. Apart from special terms referring to particular distributions (e.g. *isohyet* for rainfall) *isarithm*, *isopleth*, *isometric line* and *isoline* have all been used to describe generally a line on a map drawn through places having the same value of a certain element. People have been rude about 'isoline' describing it as 'unfortunately a bastard word' and 'a horrible hybrid' but have admitted that it 'may be *self-evident in its meaning*' (author's italics). Enough said.

from the lower value. In practice these distances are almost always judged by eye and in the above example estimates of just over one quarter and one half would suffice.

Calculations such as this establish points through which isolines should pass and the isolines themselves can be drawn in by joining up all established points relating to a particular value, but the process is not always so simple as it sounds. Common sense will often suggest some modification to the pattern which would derive simply by following the method described above for this has one major weakness: it presupposes that between any pair of spot-heights there will be found only intermediate values and never anything larger or smaller than either of them. To return to the relief analogy, it discounts the theory that either a marked ridge or valley could intervene between any pair of spot-heights. Rather complicated results may be obtained if this idea is too strictly adhered to, the most usual characteristics being the presence of many isolated 'knolls' or 'hollows' originating from isolated high or low values; another approach, producing simpler isoline patterns, would be to treat these values as indicators of tongues of higher or lower value penetrating other areas.

Even when simple modifications of this kind are observed there is still room for a good deal of personal judgement in deciding just how and where isolines should be drawn, and two cartographers working from the same distribution of values will not necessarily produce identical results using this technique. Detailed knowledge of finer aspects of the distribution on the ground should always be used (where possible) as an additional aid in drawing the isolines and may suggest modifications to the basic proportional spacing, such as closer bunching for sharp transitions or added promontories and embayments not supported by any actual values.

The repeated reference to topographical terms in a description of this technique is almost inevitable in view of its noticeable analogy to relief maps, but it should be noted that major differences *in principle and practice* may have to be accepted if the best results are to be obtained. The essential feature of contoured maps drawn from spot heights of known value is that it is certain that between the spot heights there must occur *somewhere* places which exhibit every intermediate value, even though several of these may coincide in a cliff or crag. Thus between spot heights of 60 m and 100 m there must be heights of 61 m, 62 m, 85 m, 90 m and so on. In a few cases where statistics are being represented by an isoline technique, a similar kind of distribution can be assumed. The most obvious group-example is climatic statistics for such quantities as temperature, pressure, rainfall or humidity. In these cases we are dealing with natural processes capable of infinite variation in which changes, though they may be quite sharp, cannot be abrupt; once again between places with 60 mm and 100 mm of rainfall there must be somewhere with 61 mm, 62 mm and so on. When man-made distributions are being considered this kind of gradation need not follow, and changes may be entirely abrupt with no inter-mediate values at all. Thus an average population density of 2,000 per square kilometre for an urban unit and one of 200 for an adjacent rural area will not necessarily be separated by a zone of intermediate values. Although it is common to find the highest densities in urban units at the centre with lower densities as one goes outwards, this is only a general rule and the transition from 2,000 to 200 may literally occur on either side of a hedge—buildings on one side and fields on the other. An isoline map drawn to illustrate this distribution would normally include several intermediate isolines, admittedly very closely spaced, to mark this transition.

It can be seen that we have, in effect, two rather different classes of isoline maps, one corresponding to each situation. The first type attempts to use established values as a basis for interpolation all over the face of the map, the isolines helping estimation of values at places where statistics have not been recorded. This type of map aims at a high degree of accuracy and the spacing and drawing of the isolines needs treating with care.

In the second type the isolines' function is rather different; their essential purpose is to act as a visual guide to the interpretation of trends and tendencies which the plotted values indicate. The significance of intermediate values between these known points will depend very much on the kind of distribution being plotted and especially the degree to which the 'spot-heights' used reflect typical conditions in their area. Thus in the case of land in farms in the counties of Michigan (Figures 16D and E (p. 52)) it does not seem unreasonable to assume that, throughout the southern portion of the area, the figure for the whole county is fairly representative of general conditions within that county. The author could find no overwhelming evidence of major 'unfarmed patches' of a size large enough to affect the result, small lakes and an occasional sizeable town being the only obvious, but relatively unimportant, exceptions. The picture which isolines present when drawn between spot-heights with this kind of distribution is probably fairly representative and the gradual transitions shown might be expected in reality. Unfortunately, conditions in the northern part of the peninsula are quite different; land in farms has a much more patchy distribution and the transition between farmed areas and large areas without farmed land seems fairly abrupt. The dot map in Figure 15A gives some idea of this patchiness and average figures for a whole county have much less real meaning than further south so that the rather gradual changes which the isoline map indicates are unlikely to be found in practice and both higher and lower values than the ones shown might well be expected within many areas.

When showing man-made distributions with isoline maps of this second type, there seems no reason to pursue the analogy with relief too far, and with very patchy distributions it should be quite permissible to pass from a low value to a high value abruptly, without any need to include all intermediate isolines (see, for example, Figure 46B (p. 146)). Such a modification of the method would be the most successful way of overcoming the hiatus between urban and rural densities, for example, with the added advantage that the change would take place at the most appropriate place and not, as with shading maps, at a fixed boundary line; on the other hand, the unimportant and often abrupt changes of average density within a rural area can be satisfactorily smoothed out using the normal isoline technique, giving a more flexible result than can be achieved with the ordinary shading map. With isoline maps of any type it is often useful to indicate all spot-height values from which the map was constructed so that the amount of interpolation which has been necessary can be seen at a glance.

Figures 16D and E (p. 52) show two examples of the data used in Figures 16A and B redrawn by this method. The effect of choosing different isoline values is clearly apparent, but the differential result here is due more to the various layer-colourings being spread over different ranges; in theory the isolines on one map, though omitted from the other, would occupy identical positions on both maps if they had to be drawn in.

As might be expected from the theory underlying their construction, isoline maps are best suited to illustrating distributions where changes are relatively gradual; they are particularly unsuited to distributions which are patchy or uneven, especially where the areas used as a basis are many and small. Circumstances such as this will often produce very contrived isoline patterns unless the isoline values are so widely spaced that a great many of the minor variations in the figures can be ignored. This explains the apparent ease with which layer-coloured isoline maps are often used in atlases to show such notoriously varied and irregularly distributed quantities as population density. Observation will show that each colour covers a very wide range of density, often increasing in an irregular way, e.g. by using 1, 10, 25, 100 and 200 persons per square kilometre as critical values; if *detail* has to be maintained in a patchy distribution a shading map will often allow easier construction and give better results.

Repeated quantitative symbols
Many of the methods used for showing the distribution of quantities at a series of points can be adapted for use with quantities relating to areas. As the methods have already been described it will only be necessary here to review those aspects which are particularly relevant to this new task.

The idea of representing quantity contained within an area by a single symbol has a strong appeal in many cases. As we have seen, techniques relating to areas frequently need quantities expressed in a particular way so that the map shall not be confused by the effect of the varying size of the units used; but the answer may prove to be rather unrepresentative. The whole concept of density of distribution within an area, for example, is often more theoretical than real; to say that within a given area the average density is 1,000 persons per square kilometre means nothing other than that the population in that unit is 1,000 times greater than the area, measured in square kilometres, and the figure may or may not have the remotest resemblance to conditions within the area. If this resemblance is known to be slight or doubtful it might to best to abandon such artificial concepts and stick to one symbol placed within the area and representing the actual quantity concerned.

Unfortunately the results are often much less successful than when similar ideas are applied to point values, for where figures relating to areas are used what matters is frequently not the actual quantity but the variations in distribution from area to area which it implies. It is, of course, precisely for this reason that concepts such as density are introduced at all and it must be admitted that the alternative of piecing together the same picture from a large number of varying symbols, each attached to an area also varying in size, is not very successful. Figure 15C (p. 48), which illustrates land in farms in Michigan treated by this method, brings out the disadvantages which so often result. It is always difficult to strike a balance between a suitable symbol size range and the areas within which these must be contained, e.g. in Figure 15C the rather large symbols make it impossible to add the rectangular boundary lines without causing confusion. The method is too 'rigid' and lacks any element which helps the eye to move from one part of the map to another. A more flexible quantitative symbol, such as the dot, will usually achieve a much better result. Another example, drawn in a slightly different context, will be found in Figure 38 (p. 119).

Repeated statistical diagrams

Statistical diagrams, of the kind described at the beginning of this chapter are also often encountered drawn on a map base to illustrate variations in several factors throughout an area. The method has some of the same potential and problems as repeated quantitative symbols, for once again the resultant effect is produced by several individual items scattered about the fact of the map. Its success depends on the purpose which the map is intended to serve.

In some cases the map is little more than a visual geographical index to diagrams which would be drawn in any case, and where this is so it may be sensible, if room permits, to place these on a map background which will illustrate their location and allow minor comparisons, e.g. between neighbours, to be made as well. In many other examples, however, it would appear that far more than this is required of the method and the map user is being asked to piece together from these many scattered items a picture of the variations in distribution which they indicate throughout the map. This is surely asking too much. It is difficult enough to evaluate the relationship between symbol and area and its variation over the map when quantitative symbols are being used, it is almost impossible to do this when we have not a simple geometrical shape but a complex diagram such as a graph or pie graph, which may contain several variable elements all relevant to the theme of the map.

Figure 17 illustrates, in a manner commonly employed for this purpose, the relative importance of different types of land use in the provinces of the Netherlands and the weaknesses of the method can be seen fairly easily, even on this relatively simple example. It is not difficult to obtain an impression of the importance of the different uses in any one province, nor even to compare this with its neighbours, but it *would* be to try to maintain this comparison over the whole of the map to obtain a composite picture of land use distribution throughout the kingdom; it is not even certain that the map would be a very reliable way of obtaining an impression of the varying distribution of only one element, e.g. arable land, let alone five.

The trouble here is that the cartographer has done too little work on his statistics and left the really hard job of correlation to the map user. The responsibility for evaluating the important aspects of any distribution should always rest with the cartographer who should first determine these and then illustrate them in map form, e.g. by adding an inset diagram, as has been done here. Reliance on visual comparison by the map user is no substitute for this direct mapping of relevant details. The same problem occurs where change in a situation is illustrated by mapping it at two different dates. If changes are *really* important then they should be mapped separately and not be left to a comparison of two maps.

In all fairness it should be added that it is very tempting to try to use this method of dealing with rather complex distributions, for the simple reason that it manages to kill so many birds with one stone. It is possible, for example, as in Figure 18A to try to illustrate variations in rainfall throughout the year in Australia by drawing on a map of that continent a series of small bar graphs showing mean monthly rainfall at selected stations. Without any further help this provides an impossible amount of detail for the map user to assess; an alternative might be found by constructing not one map but a series of

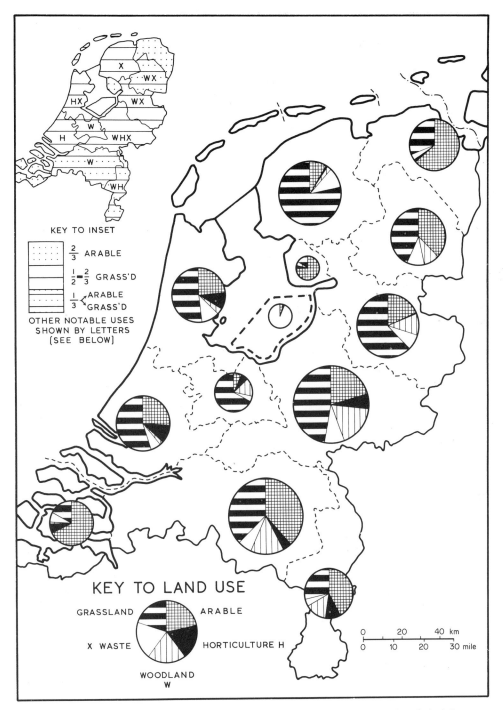

Figure 17 Land utilisation in the provinces of the Netherlands shown by repeated statistical diagrams (pie graphs). The inset map is added to summarise the situation portrayed on the main map and thus help interpretation.

Source: Central Bureau Voor de Statistik, *Jaarcijfers Voor Nederland 1957–8*, Zeist, 1960.

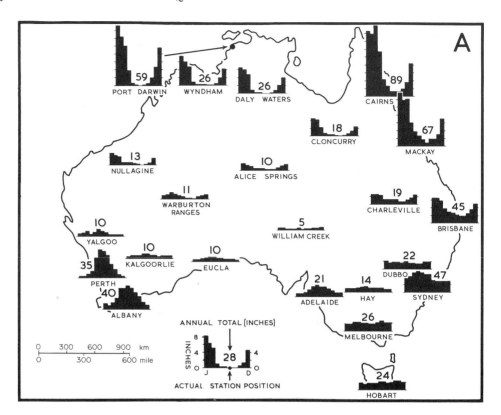

Figure 18   Distribution of rainfall in Australia shown by: **A.** Repeated statistical diagrams (bar graphs). **B** (opposite). The same method with the message of the symbols emphasised by annotation recording main themes. Most maps using repeated statistical diagrams would benefit by annotation of some kind, bringing out essential points from amongst a welter of detail.

Source: Meteorological Office, *Tables of Temperature, Relative Humidity and Precipitation for the World (Part VI),* London (H.M.S.O.), 1958.

isoline maps showing, say, mean monthly rainfall only for January, April, July and October, although this still leaves a considerable amount of comparison to be done if a composite picture for the whole yeary is required. Another possibility lies in amplifying the original method by examining the detailed statistics further, ruthlessly eliminating detail and determining any major similarities in yearly régime which exist at any of the stations shown. If these characteristic similarities are then marked on the map as a general guide, the usefulness of the individual diagrams is much increased. General trends are already selected and the diagrams can supplement these by indicating any minor variations within the major themes. Figure 18B shows how much more useful such an annotated map is than the 'bald' diagrammatic representation in Figure 18A. Other types of statistics with several variables, e.g. for land use or industrial structure, could be treated in a similar way, e.g. Figure 51 where background shading replaces annotation.

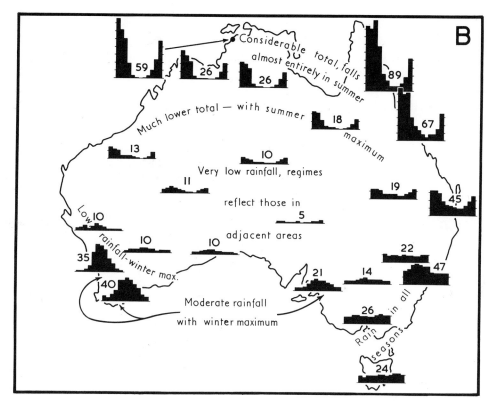

Figure 18B

Once again we have used here the idea of grouping, this time grouping around certain common descriptive headings, to sort out a complex situation and make it more easily appreciable. The trouble with mapping and appreciating distributions such as those shown in Figures 17 and 18 is that they are highly complex because they contain *multiple variables*. In Figure 17 we are trying to show the variations of 5 things (5 different land uses) all at once and in Figure 18 no less than 12 variables (12 monthly rainfall totals) have to be dealt with. This problem of adequately representing *multiple* variables on one map is perhaps the single most difficult task in the whole of statistical mapping. Almost inevitably grouping, whether by summary diagram, as in Figure 17, summary notes as in Figure 18 B, or by some more complex method will be needed; the whole question is, however, important and difficult enough to be given separate treatment elsewhere and a more elaborate and detailed description of methods available and possible solutions will be given in the second section of Chapter 4.

E QUANTITIES DISTRIBUTED ALONG A LINE

Of all the major groups of quantitative statistical maps this is by far the simplest to describe, for only one method is in common use. The quantities portrayed are almost always traffic flows along routeways of various kinds and the method used is to draw, astride each route, a band proportional in width to the quantity of traffic passing.

The idea is quite successful, but it should be noticed that the scale of band-widths is a linear one and, as with all such scales, an extended range of values cannot easily be accommodated. Where such a range has to be dealt with a convenient modification is to

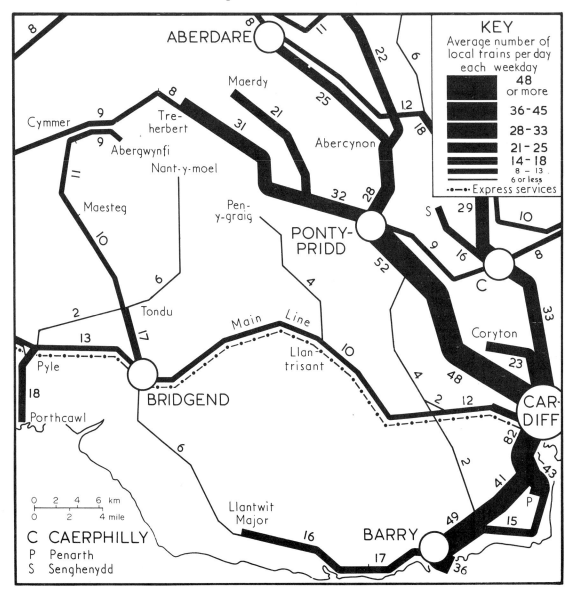

Figure 19 Local passenger trains on railways in part of south Wales, 1955, shown by bands of proportional width. *Source: Bradshaw's British Railways Guide, July, 1955, London, 1955.*

adopt a series of graduated band-widths, designed after the manner of the graduated range of symbols for point values which were described earlier. Figure 19 shows an example of such a map.

The methods described in this chapter have been taken in the order set out in Figure 2 (p. 17). As that diagram itself suggests some methods can be used in the presentation of more than one kind of distributional pattern and it will have become apparent from the text that even more cross-linkages are possible. The particular choice of one method out of the several which are available to tackle some problems will depend on many factors, and particularly on certain general considerations which will be described in Chapter 5, but before any such choice is made it is often desirable to have a look at the statistics themselves. These will form the raw material for whichever method is chosen and some of the more important aspects of their character are considered in the next chapter.

3 Sense from statistics: the search for the significant

It is remarkably easy to slip into the habit of summarising a situation statistically. We tend to say, for example, that in Bloggsbridge Urban District 55% of the women under thirty are brunettes, 40% are blondes and 5% are red-heads, in preference to saying (what amounts to the same thing) that there are rather more brunettes than blondes whilst red-heads arc quite uncommon. One great advantage in using the statistical approach in preference to the descriptive one is that it *seems* to be so much more precise and it makes comparisons with other places so much simpler. Similar figures for the neighbouring urban district of Mudworth, for example, would immediately make it apparent to those gentlemen who prefer blondes whether their efforts are more likely to come to a successful conclusion if transferred to that locality.[1]

This idea of using a series of statistics as a sort of shorthand summary of a complex situation at one place and from them making comparisons with other places is so common a procedure that it is easy to use it in rather a slipshod manner. If the method is to succeed at all one absolute prerequisite is that the statistics used to represent conditions at a place should *really* do so, and should have their roots in geographical reality and not, as so often happens, in administrative convenience or mathematical calculations, both of which are capable of producing answers which are not only inaccurate but quite unlike any set of conditions which actually exist.

We have already encountered the problem of unrepresentative values when considering techniques for presenting quantities relating to a given area and it is worth while looking rather more deeply into this general problem. Of the two main causes of unrepresentative figures mentioned above, those which derive from administrative causes are probably the most difficult to detect and the most difficult to overcome. A major source of this difficulty lies in the fact that many people who need statistical information to help them with a particular problem have to use statistics which are provided by other agencies, often for quite different purposes. Occasionally it may be possible to devise and obtain privately exactly those statistics which are needed to carry out a task but even here, to save time or expense, there is a tendency to incorporate or at least start out from, statistics obtained from other sources such as a national census. Censuses and similar publications by government departments or large-scale organisations such as UNO are invaluable to the statistical cartographer simply because they offer information in considerable detail on

[1] This is not strictly true, of course. All that comparative statistics of the above type would indicate is the probability of *encountering* ladies of a particular hair colouring. What happens after that would depend on many things, amongst which statistics of quite a different type might play a part.

a wide range of subjects; unfortunately, they almost inevitably involve certain disadvantages which are characteristic of 'other people's statistics' and a sensible awareness
of these disadvantages can do much to improve the quality of any work which derives
from this type of source material.

The fundamental principle which should be kept in mind when using official statistics
of any sort is that these are a means to an end rather than an end in itself. Official statistics
are not an implacable umpire whose pronouncements must be complied with lest they be
'used in evidence against you', but are instead a valuable source of information which
may need certain modifications before it can be used for the work in hand.

The most unsatisfactory feature which pertains to 'other people's statistics' is usually
the units or areas for which individual figures are given. These are almost always the
various types of administrative subdivision of a country or large organization (such as
British Railways or the National Coal Board), and though these may be well devised as a
basis for administering that country or organisation they may be much less meaningful
in terms of other aspects of everyday life, and statistics which relate to them may be less
useful generally than one might expect. Until recently when official statistics were being
prepared, little consideration seems to have been given to the idea that anyone might
want to map these for units which differed from official administrative divisions, despite
the fact that one might often get a much more representative picture if a distribution
was mapped in terms of some *other* type of areal unit, e.g. population density or employment mapped by grid squares. Fortunately due largely to the stimulus and potential of
computer mapping (see Chapter 8) some countries are now designing their censuses to allow
this. Even so in many cases serious problems remain causing the 'official' figures to give
no guide whatsoever to the real-life situation. Let us take a simple example.

The population of a city is often quoted to give some indication of its size and the kind
of place it is. Any foreign person who consulted the Preliminary Report of the British
Census, 1971 would find that the populations (in thousands) of the largest English cities
outside London were in 1968

Birmingham	1,013	Leeds	495
Liverpool	607	Bristol	425
Manchester	541	Coventry	335
Sheffield	520	Nottingham	300

A moment's thought and a glance at an Ordnance Survey map will soon make it
apparent that as a guide to size and 'kind of place' these figures have not a great deal of
meaning. Many British people would be able to help a less well-informed foreigner by
pointing out that Salford, with 131,000 population was simply part of the Mancunian
rose smelling as sweetly by another name, but would they also realise that something so
essentially 'Manchester' as the Trafford Park Industrial Estate is not technically so but
is situated in Stretford Municipal Borough next door? In the list above only the figures
given for Sheffield, Coventry, and Leeds begin to approach being representative of what
most people think of as these towns. Nottingham, and Newcastle to take another example,
are both surrounded by suburban 'Urban Districts', and the inclusion with Newcastle of
no more than Gateshead alters the picture completely and brings that city up from 13th

to 8th place in the list. A revised list incorporating only the most obvious additions presents quite a different picture:

Birmingham-Warley-Solihull	1,283	Bristol-Kingswood-Mangotsfield	478
Manchester-Salford-Stretford	726		
Liverpool-Bootle	681	Nottingham-Arnold-Carlton-West Bridgeford	407
Sheffield	520		
Leeds	495	Newcastle-Gateshead	317

Equally misleading are the mythical entities which occur from time to time in official figures. The gathering together of six towns,[1] none of which is more than medium sized, may produce the City of Stoke-on-Trent with 265,000 people, but this is not remotely to be compared with a large city such as Leicester (284,000), nor is the agglomeration of mining villages which make up the Rhondda Municipal Borough (89,000) anything like, say, Cambridge (99,000); yet these bare official figures are widely quoted whenever information is rapidly sought and comparisons are not infrequently based on them.[2] Whilst it is relatively easy to make allowances for discrepancies of this kind in the more familiar circumstances of one's own country it is far easier to overlook them when using foreign statistics.

Fortunately the misleading nature of 'official' populations is becoming increasingly recognised and many publications now give two figures for large cities, one for the city 'proper' and the other for 'greater X'. One has to say 'greater X' advisedly for often the exact connotation of the larger unit, and the reasons for including places within it are not defined and one suspects that there is often scarcely any better degree of standardisation in comparing figures for 'greater Xs' than for cities proper. Let an example make the point. *The Geographical Digest*, 1970 (London, 1971) gives the following figures for the 1968 populations of a sample of French cities:

Paris	8,196,746	Lille	881,439
City	2,590,771	City	190,546
Marseilles	964,412	Valenciennes	223,629
City	889,029	City	46,626
Le Havre	247,374	Dunkerque	143,525
City	199,509	City	27,504
Montpellier	171,467		
City	161,910		

The basis for the larger figures is not stated but whereas for cities in the first column the larger figures seem to be derived on a city-plus-suburb basis and relate broadly therefore

[1] *Pace* Arnold Bennett, there really are six.
[2] An interesting example of this common error was an argument appearing in the local press, that if a city such as Toronto, population 665,000 *and therefore similar in size to Leeds*, population 495,000, could support an underground transport system so could Leeds. Though the population quoted for the city was quite correct the population of 'greater' or metropolitan Toronto is in fact about 2,200,000, a totally different story when there is no equivalent 'greater' Leeds to set against it.

to a functional entity, those in the second column must surely be simply 'areal agglomerations' of the Stoke-on-Trent and Rhondda type referred to above.

Some degree of uniformity and consistency in the definition of 'greater cities' is obviously desirable here, perhaps along the lines of the well known and extremely useful Standard Metropolitan Statistical Areas defined in the 1960 Census of the United States. The concept has evolved progressively via 'Metropolitan Districts' (Censuses 1910–40) and 'Standard Metropolitan Areas' (1950 Census) but in each case the inclusion of an area was based on uniform (and often quite complicated) principles applied throughout the whole country; a description of the evolution of the idea and the basic principles of definition will be found in R. E. Murphy, *The American City*, New York, 1966, Chapter 2.

It is not surprising therefore, that the plotting of 'crude' official figures will often produce maps which are much less representative than they should be, and a sensible rule to adopt is never to plot on a map any information which is *known* to be misleading or meaningless. Quite often it is possible, with only a little research or calculation, to produce figures which, though they may result from subjective judgement and be open to criticism on the grounds of inaccuracy, are much more real and representative than the original ones.

Unfortunately, one encounters all too often maps where convenience of source material seems to have been the only selective criterion used in compiling the map. An example of these defects and their possible improvement is shown in Figure 20 which illustrates the distribution of urban population in the central West Riding of Yorkshire in 1951. Figure 20A shows this in a manner similar to that used in the 'Population of Urban Areas' map published by the Ordnance Survey and deals only in the population of the administrative units in the area; Figure 20B ignores local government boundaries and attempts to

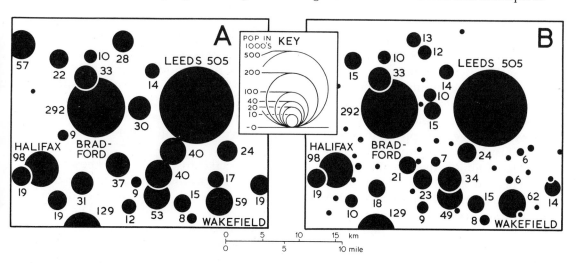

Figure 20   Population of urban areas in part of the West Riding of Yorkshire, 1951. In **A** urban areas are defined on a purely administrative basis; **B** refers to the actual physical units 'on the ground'.

Source: (Figure 20A) Registrar-General, *Census of England and Wales, 1951*, County Volume: Yorkshire (West Riding), London (H.M.S.O.), 1954.

show the actual population of the physical units which form the urban structure of the area. Within the West Riding the boundary revisions of the 1930s often grouped together former small Urban Districts based on the large industrial 'villages' (of about 5,000 to 7,000 population) and the few *small* towns which lie in the area between the large county boroughs; the resulting medium-sized 'towns' which appear on Figure 20A are unreal entities and, as Figure 20B shows, there are almost no medium-sized units of between 25,000 and 45,000 population within the area of the map. It is ironical to note that though populations in this area are generally fairly static and have been so for some time Figure 20A would have had a completely different appearance if it had been prepared before the boundary changes occurred. To be quite fair, the Ordnance Survey's map does not purport to show the population of towns in Britain but only of 'urban areas', but the latter term is an equivocal one and if the appellation 'urban' is applied to administrative units in as meaningless a way as is the case in Great Britain the picture which the map portrays will be of limited usefulness.

Not only do meaningless quantitative statistics give a false picture of a particular place, they render comparisons with other places of doubtful reliability. Suppose for example that we wish to compare regional structure around Leeds and Newcastle-uponTyne by studying the part which commuting plays in providing part of the labour force of each town. In 1966 the commuting figures for the two towns were as follows:

	Leeds	Newcastle	Newcastle and Gateshead
number of jobs	271,000	173,000	225,000
number of commuters	50,000	85,000	95,000
commuters as percentage of total jobs	18	49	42

It has already been pointed out that the City of Leeds is a relatively meaningful entity whereas the City of Newcastle is not and if Gateshead is added to Newcastle and movement from one to the other ignored we get the adjusted figures given in the third column. True, the difference between 'Newcastle' and Leeds is still quite considerable, but in view of the ring of suburban 'urban' districts which lie around Newcastle and the fact that even in Leeds one of the main concentrations of industry occupying possibly 2,500 workers lies just outside the city boundary, one is left wondering whether the difference is really so large as it appears to be and whether it does not originate largely in the peculiar administrative framework of the Newcastle area.

The administrative complications which beset official statistics are well known if not well heeded. Another much less obvious cause of poor representational quality can derive from what might be termed 'mathematical complications'. A major common defect of statistics is that taken down to their smallest unit they are too prolific, either for assimilation or even for general publication, so that they have to be summarised or given as average values for increasingly large units, these larger units being frequently administrative subdivisions such as Rural Districts, counties, provinces or regions. Many people who make use of such averages or summary figures rarely pause to consider the limitations of these strictly mathematical devices which attempt to replace a series or group of values by one significant value. The usefulness of the average, for example, depends entirely on the range and comparison of the group of values from which it was derived, and it may be either highly significant or almost meaningless according to circumstances.

The statistics which form the basis of Figure 21 provide some interesting examples of the way in which these factors can work. They relate to population change in the North Riding of Yorkshire between 1931 and 1951 and were taken from the Census of the latter date. It would be possible of course to obtain one figure to represent the change within the whole county (it is in fact $+12\cdot6\%$) but with a unit which embraces extensive moorland areas and industrial Tees-side further subdivision is obviously necessary. If the county is divided into 'rural' areas (i.e. Rural Districts) and 'urban' areas (i.e. the remainder) the figures for these sections would be $+20\cdot9\%$ and $+11\cdot0\%$ respectively. Now an increase of $+20\cdot9\%$ would be startling enough in any rural area let alone one which if often barren and remote, and this figure invites further subdivision into its constituent Rural Districts. The figures for each of these show $+20\cdot9\%$ to be quite useless as a representative value, for of the 20 constituent Districts, 11 showed an increase but 9 showed a decrease; and if we take out only the two most extreme cases (Richmond and Flaxton R.D.s) the figure for the remainder is at once reduced to $+9\cdot1\%$. Even this figure is an uneasy representative of the very varied conditions encountered, and the obvious course would be to treat each Rural District separately.

Maps such as Figure 21A which do just that are quite common in British literature and it is worth while looking to see whether we have achieved any more reasonable a solution. Sometimes we have. Aysgarth R.D. had a total change of $-15\cdot3\%$ and of its 12 parishes all but one showed a decrease and of these 8 were within 3% of the figure for the whole Rural District; here is a case where the district figure is real and meaningful. Contrast this with Northallerton R.D. where of 41 parishes 20 showed increase, 21 decrease, only 23 came within 5% of the average figure and the range ran from $+76\cdot5\%$ to -75%. In Wath R.D. an aerodrome constructed in one parish completely falsifies the picture for the whole Rural District $(+49\cdot8\%)$; without this exceptional case the other 10 parishes give a total change of $-7\cdot6\%$. The average value, or a summary figure, is often of doubtful utility because it may cover up very varying conditions and particularly because it is too much affected by extreme values, allowing them to play an important part in determining the final figure.

To remove these unrepresentative values for the large units the next step would appear to be to descend to even smaller areas, in this case the constituent parishes. We could, of course, calculate the change in each parish in a Rural District and calculate the average of these, but like all averages this would be affected by extreme values.[1] Alternatively, we could use the *median*, another mathematical device which produces a representative value for a group of figures but which is less affected by extremes. The median value of any group of figures is simply the middle value when they are arranged in ascending order, all occurrences of any repeated value being counted separately. Thus the median

[1] For Wath R. D. referred to above the figure produced in this way is $+39\cdot8\%$, entirely due to the fact that the parish with the aerodrome increased by 530%, which is quite enough to overwhelm any average. This figure of $+39\cdot8\%$ is the true average of the individual parish figures. The 'summary figure' of $+49\cdot8\%$ actually given in the Census for the R.D. as a whole is, mathematically, the *weighted* average, the contribution from each parish being weighted according to its population.

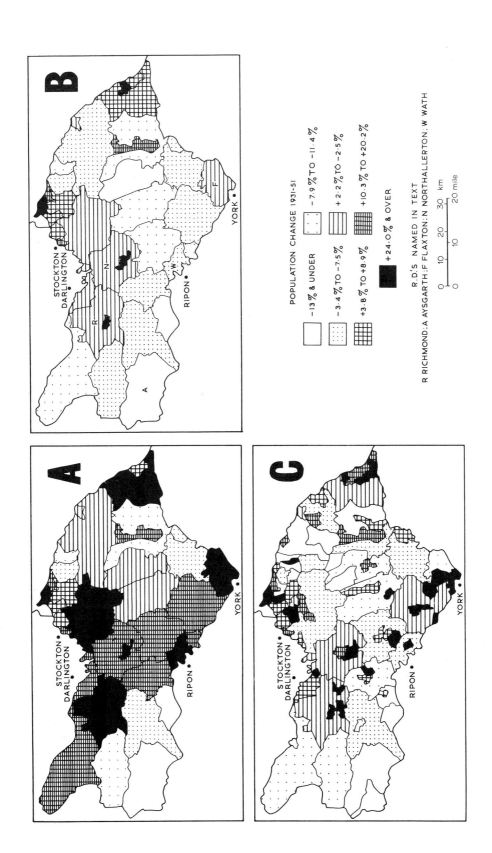

Figure 21 Population change in the North Riding of Yorkshire, 1931–51, showing the different results obtained when different types of area and unit are used as the basis of the change, namely: **A** Average change for whole Rural and Urban Districts. **B** Median of change in individual parishes in each Rural and Urban District. **C** Average change for 'rationalised' units, see text.

Source: Registrar-General, *Census of England and Wales, 1951*, County Volume, Yorkshire (East and North Ridings). London (H.M.S.O.), 1954.

of the 7 values 5, 8, 8, 13, 17, 31 and 93, is 13, the fourth or middle item on the list. If there is an even number of items so that there cannot be a middle value the convention is to give as median the quantity which lies half-way between the two values which lie astride the middle.

The great asset of the median is that exceptional and freak values in a run of figures get no more weight in the final figure than any other constituent, so that if the great majority of the constituents have normal values but there are two or three freak values in the range, the median is likely to be far more representative than the average.[1] On the other hand if the constituents themselves have a wide range generally, as in the case of Northallerton R.D. referred to above, neither the median nor the average will be very representative, and further subdivision into more homogeneous units is the only way to achieve better results.

Figures 21B and C show something of this rationalising process in action and indicate how important sensible treatment of a series of figures can be if a good result is to be obtained. Figure 21A has already been examined and shows the change when measured by the easiest available published figure—namely, that for whole administrative units. Figure 21B replaces these misleading values by the *median* value of the changes in *individual parishes* in each Rural District. The difference is startling and a good example of how the effect of a few extreme values was spread over a wide area in Figure 21A. Unfortunately, Figure 21B achieves this improvement at the expense of almost concealing the fact that there was any marked increase in rural areas; and Figure 21C attempts to rectify this by giving figures for the total change in smaller, more homogeneous units which could easily be produced by removing obvious anomalies within each Rural District. Thus all small towns which were not officially 'urban' in status but which were quite distinct from the mass of settlements in the rural area were treated separately or in conjunction with any 'overflow' into neighbouring parishes. Also treated separately were any other parishes with more than 800 people, or any with a 'remarkable' change, i.e. more than +50% or −33.3% provided the population exceeded 100 in 1931. Small detached portions of any area were not allowed but were added to the most similar adjoining large unit, and some of the peculiar Urban Districts in the north were subdivided into an urban core and rural fringe. Figure 21C naturally involved more work than Figures 21A and B, but the result was worth while. The general pattern of rural decrease stands out quite well with the 'plums in the pudding' distribution of the exceptional values seen in their true perspective. However, figures are still given for inconveniently large areas; the devices used were chosen because they were easily applied (hence the choice of +50% and −33.3%) and further possibilities for dealing with these figures will be examined in Chapter 7.

Many of the defects just described arose from trying to obtain figures for areas which were too big to have any meaning. Oddly enough it is possible to obtain somewhat parallel defects because the areas used are too small, so that statistics relating to them are too easily affected by random occurrences. Many of the rural parishes within the North Riding have ridiculously small populations; In 1951 there were 592 of them in all and

[1] The *median* of 5, 8, 8, 13, 17, 31, 93 as given above is 13, the *average* is 25 which is much less representative.

of these 297 had populations of less than 200 and the most common size groups were 50–59 population (24 parishes) and 70–79 population (25 parishes). In units as small as this, chance happenings can quite markedly affect the picture. Farmer Giles's wife producing her sixth child a couple of days before census night automatically boosts the population by 1 to 2%; the demise of the oldest inhabitant at ninety-four a month later quite fails to redeem the official statistical situation. In larger units such as urban areas happenings of this kind[1] are sufficiently frequent to cancel one another out and allow long-term trends to show through, but in small rural parishes they can quite easily cause arbitrary and misleading figures. It is often difficult to decide with small units just how large a change need be before it can be regarded as significant and not 'accidental'.

There is no easy way round the administrative and mathematical difficulties which other people's statistics so often present. The only safeguard is to know one's area, one's statistics and their limitations and to concentrate more on the figures one wants rather than be obsessed by the statistics one has got. The examples above come mostly from British figures relating to population, but there are equal pitfalls in other branches. Figures for commuting, for example, are seriously affected by the size of the unit used, so long as crossing a boundary line remains an essential definition; in the Industry Tables of the 1961 British Census employment in garages, which most people would rate as a typical service or tertiary industry, and which frequently employs 3% or more of the labour force of a town, was returned with Order VII, Vehicles, a predominantly manufacturing or secondary industrial group. It has since been transferred to a more appropriate position in the tertiary sector.

In this chapter so far, the main concern has been to find a significant value which could be used to represent a given statistical situation on a map. In another sense too, every good statistical map should illustrate this 'search for the significant' by ensuring that it emphasises the really important facts in a situation. Detail should not be allowed to dominate the map and obscure the main messages it has to show. Unfortunately, one encounters all too often maps which appear to be suffering from 'statistical indigestion', maps filled with detail amongst which the poor map user has to ferret around to sort out what matters and what does not. What are we *really* expected to deduce from the map which contains 44 pie graphs all divided into six sectors, or 72 circles all of varying sizes?—certainly we shall be lucky if we obtain an impression at all commensurate with the tremendous amount of work which went into the production of the map. This point of statistical indigestion has already been touched upon when considering repeated statistical diagrams as a possible technique. The solution suggested there and elaborated further in Chapter 4 is for the cartographer to reduce and remove some of the many variables by attempting a classification of the raw material; simpler techniques may not need such drastic treatment—the replacement of a range of proportional circles simply by one of graduated symbols will help the map user to handle and appreciate large numbers of quantities at once, and so will the addition of a small 'Summary diagram' as has been done in Figure 17 (p. 63).

More will be said of increasing the actual legibility of the various techniques in Chapter 5, but in the present context the moral is obvious. It is not good enough to say,

---

[1] Births and deaths generally, not necessarily of sixth children or nonagenarians.

needlessly,[1] 'I am working on a pretty complex statistical situation and here is a pretty complex statistical map to prove it!' If statistics are to be presented visually as an aid to understanding it is almost inevitable that one step in this process will be a *sorting out* or *simplifying* of those statistics and there seems every reason to suggest that *the person who should be expected to do this should be the cartographer not the map user.* The person drawing the map, surely, should have studied the situation, assessed its implications and therefore be most able to sort things out and offer to the map user a really stimulating visual aid which will help him to grasp the situation. With the same reasoning the map user who finds placed in front of him practically the whole of the material relevant to a statistical situation pictorially translated but otherwise virtually unsorted has just reason to complain that someone has not done his job properly. True there may be occasions, as in some of the situations with multiple variables discussed in Chapter 4, where the data are so complex that a simple map is almost impossible to produce, but even there it is emphasised that an essential aim of the techniques employed is that they should effectively sort out and group the data into more assimilable form. Making 'pie-graph pictures' of it simply will not do. Almost the only reason for presenting masses of statistics seems to be the idea of showing the reader the whole of the raw material from which the study began but when illustrations are limited in number, as they often are, this is rather a wasted opportunity. Unfortunately such maps cannot usually be used to advantage as a 'statistical quarry', either, for they frequently present the information in a way which makes it impossible to read it back quickly from the map.

There is a distinct limit to the amount which can be incorporated onto a map and appreciated by looking at it. It is by no means a bad plan not merely to ask oneself, 'What do I want this map to show?' but to have a list of priorities in this respect and be prepared to sacrifice some of the less important aims if they tend to obscure the picture.

The considerations which have been stressed in this chapter refer particularly to formal statistical maps—the finished material which accompanies presentation of a piece of work; in the early stages of research it often happens that statistical maps are being used as a *source* of ideas, rather than as happens later to *represent* ideas. At this early stage rough maps, quickly drawn from crude statistics will, and often must, suffice. It is only as work on the topic progresses that the refinements described above can be added, or the freak values and unrepresentative average adjusted as a result of further information. The main ideas of the study will also emerge and it is at this later stage that the design of the final maps should be considered. Experience with rough maps will often help too in the choice of a suitable technique for the final presentation, but other considerations which surround this important decision are discussed in Chapter 5.

Limitation of the number of illustrations which can be provided to accompany any piece of work, whether by reason of space, cost or both, may make it necessary to consider another major point concerning the statistics themselves. If it is possible to have only one map or diagram to illustrate each main distribution it becomes important to decide which *aspect* of the figures shall be presented. Even the simplest set of figures can usually be looked at from more than one point of view and where space is limited it is essential to consider which is the most relevant aspect of the problem in hand. It is not sufficient to

---

[1] Needlessly is stressed here. Some situations are so complex that inevitably they produce complex maps but far more simplification is possible than is achieved in most cases.

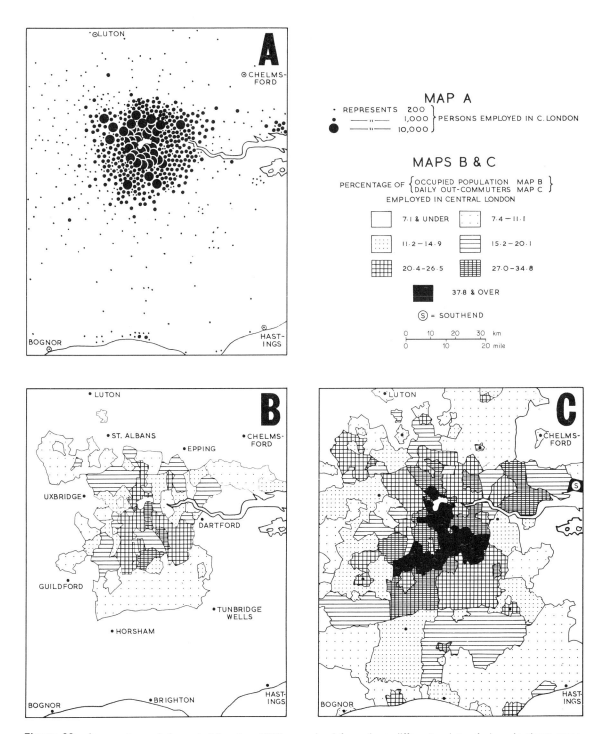

Figure 22 Journey to work in central London, 1951, examined from three different points of view. In these maps 'central London' means the cities of London and Westminster and the Metropolitan Borough of Holborn.

Source: Registrar-General, *Census of England and Wales, 1951*. Usual Residence and Workplace Volume, London (H.M.S.O.), 1956.

imagine 'I will have a map to show population distribution, or agricultural production, or commuting in an area', for example, for each of these is capable of being presented in different ways, bringing out different points and having rather dissimilar adaptations.

Consider Figure 22 which illustrates this idea for the last of the three examples given above and shows aspects of the journey to work in central London from the Home Counties area. Figure 22A is a simple quantitative map prepared by a variation of the dot method. It shows the number of daily commuters to central London but deals with this figure in isolation and in no way attempts to relate these numbers to any other aspect of life within the region; its main uses would be twofold for it offers answers to the questions, 'Where does central London's labour come from?' and, less definitely, 'How many people travel daily from X to central London?' Transport authorities would find the information presented in this way particularly useful in illustrating the amount and extent of the daily movements with which they must cope.

Figure 22B presents the same basic information in quite a different way for it deals not directly with numbers but with relative numbers or percentages. The commuters to central London are now expressed as a percentage of the total labour force in each area but no indication is given of the actual number of people involved. Instead emphasis is placed on the relative importance in each area of employment in central London, with all its implications for the student of regional association or structure and the degree of 'dormitoriness' of any particular place. To anyone involved in an attempt to plan the region as a whole considerations of this sort might be more important at first than sheer numbers.

Figure 22C is similar in principle to Figure 22B, but modifies its approach slightly. Because of the predominance of employment in service industries in Britain even the most suburban areas contain quite a large number of jobs so that the percentages in Figure 22B never rise quite so high as the stream of city-bound workers each morning might suggest. In 1951 areas such as Orpington Urban District with 27,199 workers its boundaries still had 16,221 jobs to offer them, and the 5,916 city commuters, despite their sizeable total, form only 22% of the labour force. Figure 22C attempts to counteract this considerable block of 'irreducible' home-based employment by dealing only in commuters, expressing the central London workers as a percentage of all outgoing travellers. Attention here is focused not on the part commuting plays in the local way of life, nor on the contribution each area makes to central London but on the commuter himself. It answers the question, 'What proportion of the inhabitants of X who get up and leave their home area to go to work each day are 'city-gents' and what proportion are bound elsewhere?

The division between the aspects of the figures involved in Figure 22A and Figures 22B and C is a fundamental and obvious one; that between Figures 22B and C themselves is rather more subtle, but with marked differences emerging. One of the most remarkable totals on Figure 22A is that of daily commuters from Southend[1] to London, 7,452 in all. Figure 22B shows, however, that in the light of the employment offered by a town the size of Southend this movement does not form a large percentage (11·8%) of the total

_____

[1] Marked 'S' in Figure 22C.

labour force, but in Figure 22C we can observe that if people *do* leave Southend to go to work elsewhere they tend particularly to go to central London for this purpose, this movement amounting to 38·9% of all commuters and being quite remarkable in size for a town so far from central London.

It is the ability to present different aspects of figures in this way which leads to the popular belief that 'you can prove anything with statistics' or that 'you can make statistics say anything you like'. Let us try to do just that. We will suppose that the Bloggsbridge Urban District Council have designs to incorporate within their area the Urban District of Mudworth and that one of the benefits they claim that they can offer is that Bloggsbridge provides a higher standard of urban amenities than its smaller neighbour; being thoroughly 'publicity-minded' they decide to initiate a poster campaign to influence public opinion in Mudworth and are currently devoting their attention to shopping

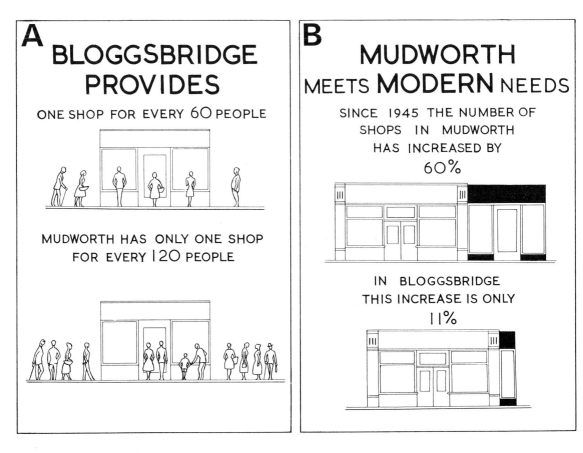

Figure 23 Two contending parties each deriving advantage from the same statistical situation by concentrating on different aspects of the figures. Notice that once one has decided to be thoroughly biased it is easy to introduce further visual overtones—the suggestion of a perpetual queue at Mudworth shops in **A** or that only shops build post-1955 are modern in **B** are examples of this.

facilities. The position is this:

Bloggsbridge and Mudworth—shopping facilities

	Bloggsbridge U.D.	Mudworth U.D.
population	24,000	12,000
number of shops, 1961	400	100
person per shop, 1961	60	120
number of shops, 1945	360	63
percentage increase in shops, 1945–61	11	60

The figures seem overwhelmingly to help Bloggsbridge, who therefore campaign in the manner of Figure 23A. The men of Mudworth know a thing or two, however. In terms of quantity, even of quality, their case is hopeless. The gloomy Victorian façades of the Mudworth Co-operative Society and 'Manchester House' or the bizarre advertisements outside 'Joe's Murder Shop'[1] are poor assets to set against Bloggsbridge's Woolworth's, Marks and Spencer's or the emporium of Joshua Bloggs Ltd.;[2] they must find some other aspect of the figures if they are to succeed in counteracting the impression put forward by their neighbours. Their answer lies, as Figure 23B shows, in quite another approach—namely, a consideration of what has been done recently. Mudworth, like many another small industrial town, has prospered unusually in recent years and the new shops added in this phase provide just the ammunition which is needed.

Their case is in many ways already lost and is based on a statistical quibble but it illustrates the remarkable effect which choice of aspect can have in depicting a situation and the danger of accepting statistical values without as full a knowledge as possible of their background. All too often figures illustrating only one narrow aspect of a complicated situation are presented with great confidence and assurance as providing the last incontrovertible word on any particular point. It is essential, so long as statistics continue to be presented to the public in support of policies and points of view, that their limitations as well as their advantages should be recognised.

[1] 'We don't slash prices—we *murder* 'em!'
[2] 'Buy it at Bloggs'!' (registered trade mark).

4 More sense from statistics: sorting-out processes—grouping and dealing with several variables

In several of the techniques described in Chapter 2 an essential part of the process was the division of a mass of data into groups which shared common characteristics and which therefore could be appropriately represented by a particular symbol, shading or annotated phrase (as in Figure 18B). The importance of this grouping process was stressed in Chapter 2 and was further touched upon in Chapter 3. The purpose of this chapter is to consider ways in which successful grouping can be achieved, beginning with situations where only one variable is involved and passing on to the much more complex situations where several variables occur.

4.1 Grouping techniques involving a single variable

The commonest kind of situation where a mass of data requires grouping is one in which only one variable is involved. Even so the number of possible techniques is quite large, at least half-a-dozen being in common use. Since a prerequisite for the selection of the best method is a knowledge of the properties of all possible alternatives, the principles and 'mechanics' only of each of the methods will be described first, an evaluation of their relative advantages and disadvantages being taken up later when all have been considered. For illustrative purposes, and to enable sensible comparisons to be made between them, each method has been used in turn to prepare a shading map from the same set of data—rural population densities for 105 minor civil divisions of 7 counties in Central Kansas, U.S.A.[1] The range of values to be dealt with runs from 1·6 to 104·4 persons per square mile, though it will be seen later (Figure 25B) that 101 of these 105 areas have values lying between 1·6 and 13·5 persons per square mile. The main methods or groups of methods available for grouping are as follows:

(a) '*Simple*' *division techniques* such as round numbers or regular intervals. These involve dividing the range of values arbitrarily. One way is to choose 'convenient' round numbers for division points, the selection depending upon the range of values involved. In this case, as illustrated in Figure 24A, 5, 10, 20, 50 and 100 have been selected and would give six classes, which is quite a suitable number for a shading map.

[1] This data was first used for a very similar purpose by G. F. Jenks in 'Generalisations in Statistical Mapping', *Annals of the Association of American Geographers*, Vol. 53, March 1963, pp. 15–26. Its presentation here in Figure 24 is derived and adapted from ideas in Figures 2–12 in Jenks's original paper, where even further examples of possible ways of grouping and presentation will be found.

Another way is to decide the number of groups which are needed and then to divide the range exactly or 'approximately' into that number of equal divisions. Suppose that 6 groups are decided upon. The range of values here is $103\cdot4 - 1\cdot6 = 101\cdot8$; precise division of this into 6 gives $101\cdot8 \div 6 = 16\cdot97$ and units of $16\cdot97$ are added on to the lowest value to give the division points, thus $1\cdot6 + 16\cdot97 = 18\cdot57(18\cdot6)$; $18\cdot57 + 16\cdot97 = 35\cdot54(35\cdot5)$; $35\cdot54 + 16\cdot97 = 52\cdot51(52\cdot5)$ and so on. A simpler approximate method is to choose a number slightly greater than the range but which is exactly divisible, and work from this. In this case 102 (the range is $101\cdot8$) gives $102 \div 6 = 17$ and division points of $1\cdot6 + 17 = 18\cdot6$, $35\cdot6$, $52\cdot6$ and so on result. These are the values used in Figure 24B.

(b) *Quantiles.* In Chapter 3 (p. 73) we met the *median* a value which divides a set of *ranked* numbers into *two* groups both containing *the same number of items*. If in a similar way we divide a set of numbers into 3 groups with the same number of items in each group the two division points are called the upper and lower *terciles*; for 4 groups we get three division points called the upper *quartile*, median,[1] and lower *quartile* respectively. Values which divide a set of ranked numbers into smaller groups each containing the same number of items are collectively called *quantiles*[2] and these are sometimes used as a basis for grouping in statistical mapping. The idea is simpler in theory than practice; consider the Kansas figures for example. A set of 105 items will divide easily into 7 groups of 15, or 5 groups of 21 and either of these would be reasonable in this case. If 7 15s were used the lowest 15 values which range from $1\cdot6$ to $3\cdot3$[3] form the first group. The second group, formed by the 16th to the 30th values runs from $3\cdot4$ to $4\cdot8$ and so on; division points between the groups can be conventionally set at $3\cdot35$ etc. However there are often reasons (see below) for wanting 4, 6 or 8 groups and where the total number of items will not divide easily the odd values must be spread among the groups as seems most suitable, regularly if possible but irregularly if this is necessary to prevent repeated examples of the same values coming in two different groups. Thus in Figure 24C 6 groups have been used and since $105 \div 6 = 17$ with 3 remainder 6 groups of 18, 17, 18, 17, 18 and 17 items have been formed. Eight groups would have given $105 \div 8 = 13$ with 1 remainder giving groups of 13, 14, 13, 13, 13, 13, 13 and 13 items.

(c) *Standard Deviations.* Another matter considered along with the median in Chapter 3 was the limitations of the average (or *mean* as it is more properly called) in providing a 'summarising value' for a set of values. For example the two sets of 5 values 45, 51, 55, 55, and 64 and 25, 41, 55, 65 and 84 both have the same mean, 54, but while 54 is reasonably *representative* of values in the first set it is far less so for those in the second, two of which are numbers quite unlike 54. What is needed here to supplement the mean is another measure to indicate the *spread of values about the mean* in any set; such a measure is the

[1] If all four groups are equal the upper and lower *pairs* of groups will contain exactly the same number of items; the division point between them is therefore, by definition, the median.
[2] The other terms for groups in common use are quintiles (5 groups), sextiles (6), septiles (7), octiles (8), noniles (9), deciles (10) and percentiles (100).
[3] Remember that as with medians if any value occurs more than once each occurrence must be counted separately.

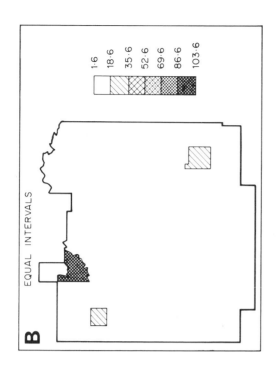

A ROUND NUMBERS

	0
	5
	10
	20
	50
	100
	104

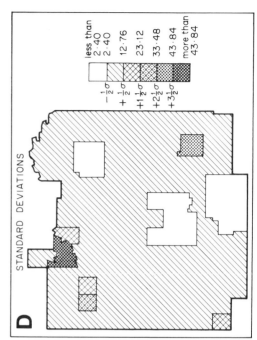

B EQUAL INTERVALS

	1·6
	18·6
	35·6
	52·6
	69·6
	86·6
	103·6

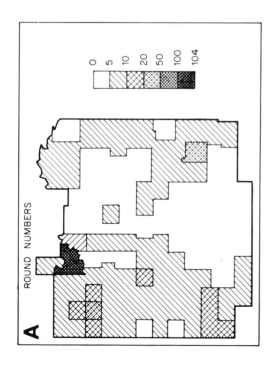

C SEXTILES

	1·6
	3·35
	4·90
	6·35
	7·15
	8·65
	103·4

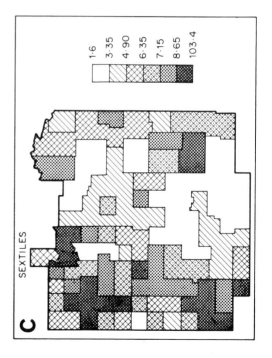

D STANDARD DEVIATIONS

	less than 2·40	
	2·40	$-\frac{1}{2}\sigma$
	12·76	$+\frac{1}{2}\sigma$
	23·12	$+1\frac{1}{2}\sigma$
	33·48	$+2\frac{1}{2}\sigma$
	43·84	$+3\frac{1}{2}\sigma$
	more than 43·84	

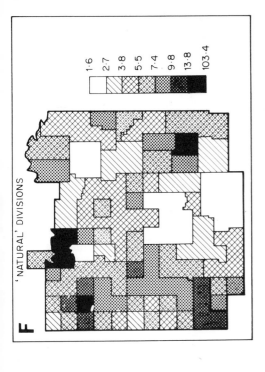

E GEOMETRIC INTERVALS

1·60
3·21
6·42
12·86
25·77
51·62
103·40

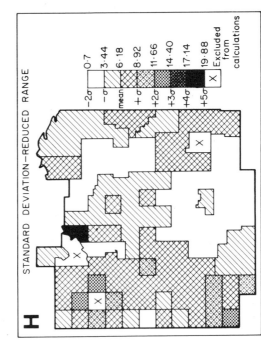

F 'NATURAL' DIVISIONS

1·6
2·7
3·8
5·5
7·4
9·8
13·8
103·4

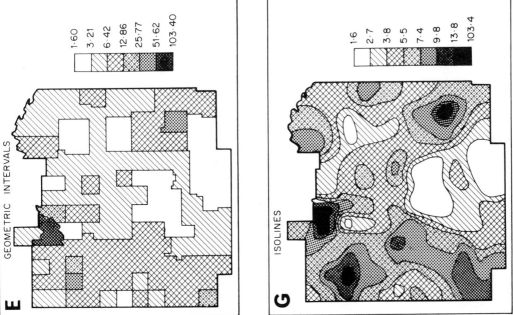

G ISOLINES

1·6
2·7
3·8
5·5
7·4
9·8
13·8
103·4

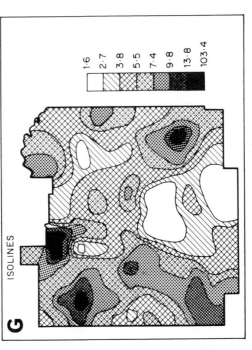

H STANDARD DEVIATION–REDUCED RANGE

0·7   −2σ
3·44   −σ
6·18   mean
8·92   +σ
11·66 +2σ
14·40 +3σ
17·14 +4σ
19·88 +5σ
X   Excluded from calculations

Figure 24 Seven shading maps and one isoline map of Rural Population Density in 105 minor civil divisions of seven counties in Central Kansas, U.S.A., showing the effect of seven possible different methods of grouping (after Jenks).

Source: Jenks, *op. cit.*

standard deviation. For readers who are not acquainted with elementary statistics the standard deviation of a set of values can be thought of as a 'kind-of-average' of the differences between values in a set of numbers and the mean of that set. It is calculated as follows:

1   work out the mean of the set of values in question
2   take the mean from each value in turn to find the *difference* between that value and the mean. *Square* this difference
3   add up all these squared differences and divide the sum by the number of items in the set
4   find the *square root* of the answer from 3 above. This is the standard deviation.[1]

Once the standard deviation and mean are calculated, groupings and division points within the set are usually established either by working outwards from the mean in units (or *class intervals* as they are called) of one standard deviation at a time, or by having a group one standard deviation wide, straddled evenly about the mean and working outwards from the limits of this group in class intervals of one standard deviation. The decision which of these two methods to adopt will often be suggested by the figures themselves. Provided that there are many values close to the mean it is always useful to define a 'near average' group and the 'straddling' procedure works well in such cases. In this particular example there are other promptings. The 105 values used give a mean of 7·58 but a standard deviation of 10·36 (caused largely by the effect of a few very high values in the group); working outwards in units of 10·36 from 7·58 would place 101 of the 105 areas in two groups which offers only very coarse differentiation. The first 10·36 has therefore been straddled about the mean to give limits of 2·40 (7·58 − 5·18) and, 12·76 (7·58 + 5·18); working outwards from these gives 23·12 (12·76 + 10·36); 33·48 (23·12 + 10·36) and so on (see Figure 24D). If figures are particularly closely bunched it may sometimes be sensible to use half a standard deviation as a class interval.

(d)   *Other mathematical intervals* e.g. geometric progressions. The division of a very large range of values into a limited number of groups can sometimes be achieved by allowing class intervals to increase regularly according to some constant rule, e.g. making the division point at the upper limit of a class always a given number of times larger than the lower limit. This is called a *geometric progression* as for example where division points 4, 16, 64 and 256 persons per square mile are used on a shading or isoline map of world population density, each limit being 4 times the preceding one. In this method one can either find suitable progressions by trial and error (in the Kansas figures using 2, 4, 8, 16, 32 and 64 would give 7 groups; 3, 9, 27 and 81 would give 5) or one can tailor-make a geometric division for the figures concerned, though it is necessary to decide in advance how many groups are needed. Let us try it with 6, the same number of groups as we have

---

[1] The method is described here in general terms for readers with no knowledge of statistics, though it seems unlikely that such readers would ever use this particular method in practice. Readers with even a slight acquaintance with statistics and who might be more inclined to use it, will know that there are quicker ways than this of calculating the standard deviation; these will be found described in any elementary textbook of statistics. The symbol σ (Greek lower case sigma) is often used for the standard deviation (see Figure 24D, key).

already found resulted in some other methods. Proceed as follows:

1   take the logarithm of the smallest value from that of the largest value and divide the difference by the number of classes. Thus log 103·4 = 2·01452; log 1·6 = 0·20412; difference = 1·81040; 1·81040 ÷ 6 = 0·30173

2   starting with the logarithm of the largest number take the final answer in 1 away from this six times (there are 6 groups). Note the answer each time. Thus 2·01452 − 0·30173 = 1·71279; 1·71279 − 0·30173 = 1·41106 and so on; check that the sixth answer is the log of the smallest value as in 1 (above)

3   look up the antilogs of the answers in 2 (above). These are the division points. Thus antilog 1·71279 = 51·62; antilog 1·41106 = 25·77; other values which would result are 12·86, 6·42, 3·21 and 1·6. Here the geometric rule produces upper limits which are always 2·0032 (actually the antilog of 0·30173 in 1 above) times the lower limits and these values have been used in Figure 24E.

Other solutions of a similar mathematical type are possible[1] but for reasons which will become apparent later they will not be discussed here.

(e) '*Natural*' *Divisions*, using a dispersal graph. A common feature of all the methods described so far is that they *impose* division points on a set of values, either arbitrarily or by using certain mathematical ideas. A totally different approach allows the figures themselves to suggest suitable division points and for this it is necessary to prepare a very simple device known as a *dispersal graph*.

On a strip of graph paper rule a scale line long enough to cover the whole range of values. Each value in turn is then plotted on this scale either as a dot or a line of unit length, any value which occurs more than once having a second dot or line added to the first, and so on. Figures 25A and B show two such dispersal graphs, the first for the values used in Figure 16, the second for the Kansas population figures referred to above and illustrated in Figure 24. Whilst the second of these graphs shows the actual values the points on the first have been rounded to the nearest whole number but it may be convenient to round off numbers to the nearest 10 or 100 with larger values. As can be seen in Figure 25 the resultant graphs show clusters or 'mounds' of dots separated by 'hollows' or breaks of varying clarity, the clusters suggesting obvious groupings and the largest breaks the most obvious place for the divisions between groups. The result defines a 'natural' subdivision of the whole range of values.

A dispersal graph does not always provide an obvious or simple solution, however. Natural breaks may be either few and indistinct or far more numerous than the number of groups which is required (a maximum of about 8 seems wise if visual distinction is to be maintained on the final map). In Figure 25A there are strong breaks at around 84, 63, 52 and 18, and a weaker one at about 74 but between 52 and 18 the position is fairly indeterminate. In this case despite two rather weak breaks at 40 and 33 the value of 35 was selected; it violates no sharp natural group, yet gives a more regular class interval and more even distribution within groups than do either of the other two. These are the values used as division points in Figures 16B, C and E. The distribution in Figure 25A

---

[1] Two, arithmetic and reciprocal, are discussed in G. F. Jenks, *op. cit.*

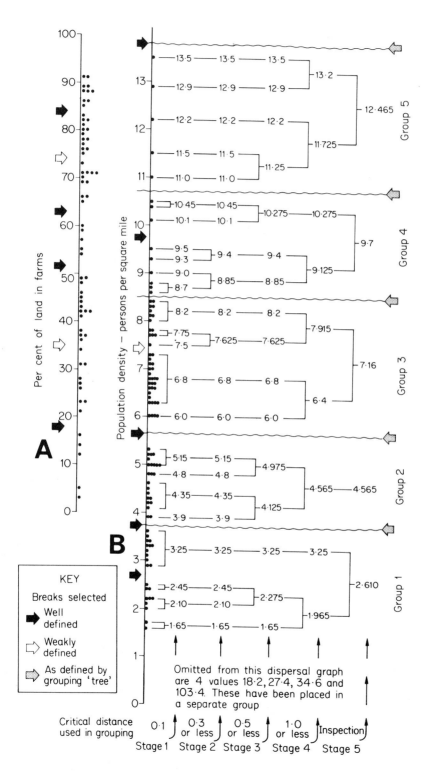

Figure 25 Two Dispersal Graphs. These two dispersal graphs indicate the spread of values for areas plotted in Figure 16 (Figure 25A) and Figure 24 (Figure 25B).

Source: As Figures 16 and 24.

illustrates well the folly of arbitrarily choosing round numbers as division points; 80 in particular cuts a well defined natural group into 2 portions and values around 70 and 90 are also best left undivided.

When selection of breaks by eye proves too difficult a *grouping tree*, as in Figure 25B will sometimes help. On the dispersal graph find the smallest interval between adjacent points (in this case 0·1 units) and amalgamate these into groups, giving to each group a value mid way between its upper and lower limits. Use these new values hereafter. Next select a slightly larger interval and amalgamate in the same way any groups or points separated by less than this value. Repeat the process until the number of groups is reduced to a suitable total. In Figure 25B 5 stages were used to produce eventually 6 groups (including the values above 13·5), the critical distances used at each stage being shown at the foot of the graph. Wherever a group or point has 2 neighbours both less than the critical distance away always amalgamate *the smallest gap first*.

The process behind the grouping tree is a relatively crude one[1] and too much reliance should not be placed on it. In Figure 25B the breaks selected by eye and used in Figure 24F seem on the whole to give better results but a grouping tree may help in less easily determined cases.

(f) *Relative advantages and disadvantages of the various methods.* Given the available range of techniques just described, which should one choose in a given situation? Which is 'best'?

There is no easy answer to this question but in the author's mind at least there is a guiding principle to follow—and it is this. Consider above all others the possibility of using natural divisions. If an area is represented on a map by a particular type of shading or symbol this suggests, whether we like it or not, similarity with other places so marked and, even more important, distinctiveness from places differently marked. If however division points have been chosen which cut arbitrarily through the figures, slicing through clusters on a dispersal graph and avoiding the breaks, the map will suggest distinctions and affinities which do not exist and which may be quite the opposite of the true situation. If we use 6 shadings or symbols on a map to represent 6 groups of things it seems most desirable that they really should indicate the 6 most natural groupings possible within the values concerned; statistical mapping is a visual technique and it is important that the method chosen should be decided upon as much for the visual consequences of its application as for the numerical properties underlying its rules.

The selection of natural breaks is therefore strongly desirable; it is not however inevitable for there are certain circumstances which may prompt the use of other seemingly more arbitrary approaches. The use of round numbers chosen arbitrarily simply to save further work must be deplored most strongly: if however one has a continuous spread of values with no marked breaks or bunches (a comparatively rare occurrence), round numbers give division points which are easily remembered and which may make the message of the map easier to remember. The choice of regular class intervals may impose on a range of values divisions no less arbitrary than those of round numbers—as in

---

[1] More subtle ideas would be to work with the true average of each group but this would be tedious to calculate.

Figure 24B where 3 of the 6 groups are blank—yet if the figures being used have not only to be mapped but also manipulated statistically e.g. for the calculation of such features as correlation coefficients, or regression lines, regular class intervals are much to be preferred. One may therefore have to accept regular class intervals in mapping to make use of them elsewhere. Conversely the irregular class intervals of natural grouping may seriously hinder statistical processing elsewhere.

As opposed to equal class intervals quantiles offer equal numbers within groups, an advantage that is often less useful than it seems to be. Particularly where there is very marked bunching of values some classes may cover only a very small range (without necessarily giving a natural group) whilst elsewhere one class may be stretched over an impossibly wide range of values, as in Figure 24C where everything over 8·65 has to go in one class to make up the necessary 17 items. However, even numbers of quantiles do offer possibilities of further useful generalisation; by amalgamation of classes one can locate the whereabouts of the top third, 'most average' third, 'most average' half of a distribution, and so on.

The single virtue of the more complex mathematical division techniques, such as geometric progression,[1] is their ability to deal with a wide range of values, but it is a virtue dearly bought, for one never knows how the results will 'fit' the distribution being plotted. In Figure 24E almost half the values fall in one division (6·42 to 12·86), yet a similar number below 6·42 is divided into 2 groups and the remaining 4 groups contain only 6 values in all, for no better reason than that the mathematics of the method demands it. Even with large ranges natural divisions can probably produce better results than this.

Unlike the other 'mathematical' approaches the advantage of using standard deviations are both more real and more subtle. The standard deviation is a very commonly used statistical measure and does truly develop from and summarise the values concerned. Because it is a *standardising measure* it is often used to compare values relating to different features at the same place, e.g. one can say that in area X rainfall is one standard deviation above the mean for that region but crop yield is 2 standard deviations above the mean, and so on. In the same way it is claimed that a series of maps of the same area can be more easily compared if all values on them are classified in terms of standard deviations, though as we shall see such comparisons are rather unreliable ways of eliciting important correlations. Like the mean the standard deviation is greatly affected by extremes and where a range of values is highly *skewed*, i.e. contains a 'tail' of very high or very low values, these may make the standard deviations a poor representative of the main mass of values. This is what happened in Figure 24D. With mean 7·58 and standard deviation 10·36, 92 out of 105 values fall in one class (2·40 to 12·76); taking out only the highest value (103·4) reduces the mean and standard deviation to 6·65 and 4·39, removing the top 3 values (27·4, 34·6 and 103·4) gives 6·18 and 2·74 respectively. It is these values, i.e. *with the top 3 areas excluded from the map*, that form the basis of the revised version shown in Figure 24H. There is good justification for doing this. Although these figures generally relate to rural population densities, the areas with the 3 highest values contain the 3 largest towns in the area, their densities are more urban than rural in their origins and

---

[1] See G. F. Jenks, *op. cit.*, for others such as arithmetic progression and reciprocals.

they are better excluded altogether, rather than being allowed to distort the whole grouping.

Figure 24H is a much better map than Figure 24D and, rather fortuitously, its groups do not cut badly across natural grouping (see Figure 25B). Yet it is the author's contention that Figure 24F still offers the better answer. There is an issue here which is really very important and which will become relevant elsewhere in this book. It is this. *Are we concerned to map reality or mathematical abstractions of reality?* It is not always an easy issue to decide. If we wish to pass on beyond mapping to statistical manipulation we may have to accept the latter, for there is often no chance of having both. Then too mathematical methods appeal because they can be applied 'impartially' to all kinds of figures and their simple underlying rules also allow easy application by computer in computer mapping (see Chapter 8), yet these are not always overriding considerations. 'Mathematical' methods are not 'respectable' or fool-proof just because they are mathematical and above all there is no inherent virtue in settling for mathematical abstractions of reality if reality itself will do equally well.

Figure 24 has been provided to give a practical illustration of the validity of the arguments put forward in the preceding pages. In studying it the reader is asked to bear in mind that the difference between the maps lies not simply between the different shapes they contain (and which catch the eye), but also in the range of *values* applied to these shapes and the resemblance these values have to reality. Shading maps always present a *stepped statistical surface* i.e. one where 'plateaux' of given value give way abruptly at 'cliffs' to other values, and this makes the effect of grouping very important. Had isolines been used the effect of grouping would be, in theory, less important. Isolines present a continuous *smoothed statistical surface* (a rolling 'landscape' with 'hills' and 'valleys'; compare Figures 24F and G) and grouping should not alter the shape of that surface but simply change the values of the isolines which are drawn in on it. In practice however we tend to read an isoline map as much by its shading designation (emphasising similarities and differences) as by its isoline patterns so that even here isolines have strong visual consequences and it is important to get the grouping right.

4.2 Mapping situations where 2 or more variables are involved

Very often in real life one encounters situations where the entity or topic under study is a complex and composite one resulting from the interplay of components or factors, each of which is itself infinitely variable. This is just the type of difficult statistical situation where one might hope that a visual aid such as statistical mapping might make a real contribution in understanding what is going on; it is particularly unfortunate, therefore, that in no other major section of statistical presentation have the results of both amateur and professional alike been so abyssmally bad and generally useless. Let us begin by examining traditional ways of dealing with this problem and their shortcomings and then try to find more effective alternatives.

Traditional ways of dealing with this problem have essentially relied on one of two methods. The first of these—much beloved by compilers of national atlases, official planning departments and the like—is to provide a *series* of maps each representing a different aspect of the area. Any coincidence, interaction, or lack of it, between distributions, is then (supposedly) deduced by comparing patterns on relevant maps. The method is typically applied where the different aspects involved are measured in different units, e.g. in an urban environmental study one may meet mortality rates, population densities, income per head, and so on.

Unfortunately it is extremely doubtful in practice whether visual comparison of maps can produce recognition of any but the most extreme and obvious coincidences. We shall meet in Chapter 5 another example of the folly of relying on visual assessments in statistical mapping but more directly relevant is a study by McCarty and Salisbury[1] which comes to broadly similar conclusions. After an extensive series of tests in which subjects were asked to estimate the degree of similarity between distributions mapped in several versions of the isoline technique they conclude that 'the findings do not support the contention that visual comparison of isopleth (isoline) maps[2] provide an effective means of determining or demonstrating the degree of association that exists between spatially distributed phenomena. Only in cases in which the degree of association is very high does the process (visual comparison) produce results which approach the standard of accuracy generally demanded in present-day research and teaching'.[3]

The second traditional method, used when many variables form component parts of a whole, e.g. land use or employment divided into several categories, has been to present these in the form of repeated statistical diagrams—graphs, bar graphs, pie graphs etc., as discussed on pages 62–65 and in Figures 17 and 18. The weakness of this approach has already been emphasised here and in Chapter 3. It fails *because it leaves the map user to do all the work* of finding similarities, comparable combinations etc., and this is not good enough. If the interaction of the spatial patterns of two or more variables is significant and important then it is the *cartographer's* job to design a map which will *show this directly by classifying the differences and similarities which occur between areas or places in terms of several variables at the same time.*

A   METHODS OF DEALING WITH COMPOSITE ASPECTS OF
       2 OR 3 VARIABLES

The two basic devices which can be used to assist in clarifying data in terms of 2 or 3 variables at the same time are the *scatter graph* and *triangular graph*, already described in Chapter 2 (pp. 27–30 and 35–37), where it was pointed out that points which are close together on these graphs are *similar with respect to both or all of the variables involved.* Here then, in the natural groups found in such graphs, is the basis of a compound classification; all that needs to be done is to divide up the points on the graphs into groups or

---

[1] H. H. McCarty and N. E. Salisbury, 'Visual Comparison of Isopleth Maps as a Means of Determining Correlations between Spatially Distributed Phenomena', Department of Geography, State University of Iowa, Iowa City, 1961.

[2] There is no reason to think that other types of maps, e.g. shading, would give different conclusions (author).

[3] McCarty and Salisbury, *op. cit.*, p. 78.

clusters, define the characteristics of each group, give to it a distinctive symbol or shading and plot its distribution on a suitable base map.

Unfortunately what is simple in theory is highly complex in practice; perhaps a study of the data used in preparing Figure 26B will give some idea of the problems encountered in tackling even a relatively simple set of values involving only 2 variables. Cursory examination of figures elsewhere for percentage population change (1950–60), and percentage negro (1960) in counties in this area of the Eastern U.S.A., had already suggested that high values in one variable were likely to be mirrored by relatively low values in the other. It was therefore decided to examine this possible interrelationship further with the ultimate aim of producing a cartographic description of the spatial situation, as in Figure 26B. The first step was simple enough, namely the plotting of the data in scatter graph form, each point being given a number which identified a particular county in the area under study (Figure 26A). It was with the second step—definition of groups of points on this graph—that the difficulties began. The definition into groups must be made by eye and though natural groupings may suggest themselves at first glance (e.g. the natural groups defined by the dotted 'circles' on Figure 26A), for descriptive purposes the boundaries of the group *will ultimately have to be drawn using lines parallel to the axes of the graph* (the pecked lines in Figure 26A). Furthermore for reasons of ease of description and representation on the map a maximum of about 10 groups seems necessary, so that grouping has inevitably to be made in rather 'broad' or 'coarse' terms; attempts to define very narrow, limited groups will soon produce far too many to be handled satisfactorily. With this in mind let us look for some obvious division lines in Figure 26A. What would be most desirable are lines which can run completely across the graph without cutting any 'natural' groups and in this case we are lucky. The lines PQ and RS on Figure 26A do just this. Elsewhere things are not so convenient, the line XY ($+7\%$ population change) for example, nearly forms the lower boundary of group B_2 and the upper boundary of group C_1, except that it severs counties 14, 16 and 61 from groups to which they obviously belong. However as we shall see, so convenient is it, in descriptive terms, to use such lines that it becomes desirable to introduce the idea of *tolerance limits*, i.e. small permissible deviations which allow a point to be included in any defined group even though it lies beyond the limits of that group. Needless to say it is desirable that tolerance limits should be small, their amount stated on the face of the map and, where feasible, an indication of their use in any area on the map marked by a symbol such as an asterisk.

Even with tolerance limits to help other problems will still remain. Natural groups may occur, but be so large that too great a variety occurs within the group. Further division is the only answer here and Groups B and C in Figure 26A have been divided into B_1 and 2 and C_1 and 2 for this reason. Then there will be other areas on the graph whose members are 'residuals' rather than a natural group, e.g. groups B_1+ and C_1+ in Figure 26A. Only rarely will a completely satisfactory answer to the grouping problem be found and it does not necessarily follow that two people working with the same figures would produce the same results; there is considerable scope here for individuality of approach and treatment. An important reason for maintaining a 'broad' or 'tolerant' approach to the grouping process above is that only in this way will one prevent intolerable

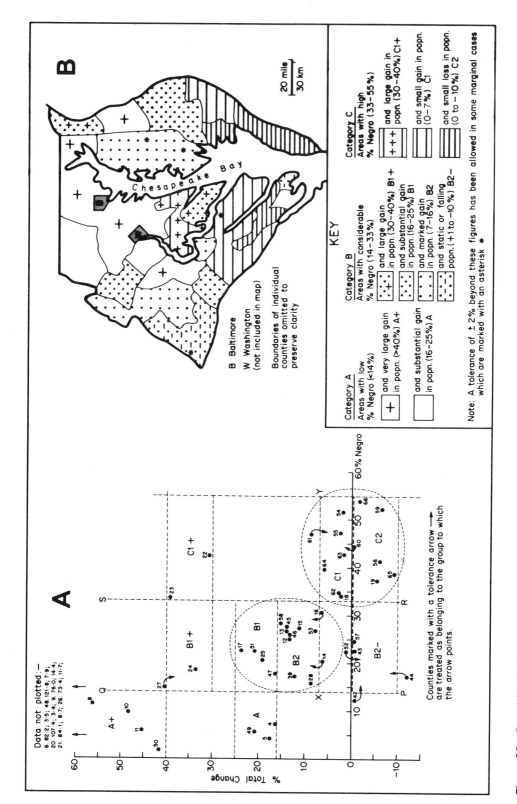

Figure 26 Counties in Delaware and parts of Maryland and West Virginia, U.S.A., classified with respect to two variables, percentage negro (1960) and percentage total change in population (1950–60).

Source: U.S. Census 1960.

difficulties arising at the next stage, that of *describing the groups*. It is, of course, perfectly possible to provide a completely numerical description, for example the groups in Figure 26A could appear as described in the key for Figure 26B as:

group	A+	A	B1+	B1	B2	B2−	C1	C2	C1+
% negro	<14%	<14%	14 to 33%	14 to 33%	14 to 33%	14 to 33%	33 to 54%	33 to 54%	33 to 54%
% pop'n	40%	16%	30%	16%	7%	0%	0%	0%	30%
change	to 120%	to 25%	to 40%	to 25%	to 16%	to −10%	to 7%	to −7%	to 40%

but it is the author's contention that in this form the material presented is rather un-assimilable to the reader. If the map user is to get the message of the map (and this after all is why it is being prepared) better results will follow from the use of general *descriptive* terms, *supported* by numbers if desired. In Figure 26B terms like 'large', 'substantial', 'marked' and 'small' have been used and by keeping boundary lines common to many groups, each has a consistent numerical meaning, though this is not essential. Assimilation of the map's message will also be helped by logical arrangement of the groups, e.g. by keeping a dominant or consistent theme in the description. Thus the 9 groups in Figure 26B have been further arranged in 3 categories depending on the percentage negro they contain; further subdivision within these groups relates to population change.

Yet another means of aiding assimilation of the map's complicated message lies in the design of suitable shadings and symbols. The three-fold division mentioned above is reflected on the map by the use of 'blank', 'dotted' and 'lined' shadings for A, B, and C categories, together with the incorporation of the + or − symbols to describe some variations. Letter symbols (see Figure 28) can also help in making more 'legible' representation of categories.

Three variables are, of course, likely to produce more problems than two, but these are not insuperable, as a glance at Figure 27 will show. In this case plotting data relating to a three-fold classification of livestock on a triangular graph produced only very vague, extremely diffuse, natural groups and where this occurs the best line of approach is to base division instead on some dominant and describable theme, bearing in mind the implications of position on triangular graphs as illustrated in Figures 11B, C, and D (p. 37). In Figure 27A, therefore, most of the points have been primarily classified in terms of percentage cattle, since this is almost always the most important element, though the groups so formed have then been subdivided (using the bisector of the angle) according to the secondary importance of sheep or 'others' (almost entirely pigs and poultry). In this case the use of 3% tolerance limits has allowed round figures to be used as dividing points and most of the dividing lines are parallel to the axes of the graph. There is, however, room for experiment here. It is not difficult to find on the graph dividing lines representing situations where A = twice B and so on which, though not employed here may be useful in other cases. Examples of these have been added *for illustrative purposes only* in Figure 27A.

The result, shown in Figure 27B is a grouping which, whilst far from obvious or ideal, is not irrational and above all is *eminently describable*. The fact that alternative groupings would give different answers is not so serious as it seems. Maps such as Figure 27B often show typical transitions, e.g. group 4 areas are often surrounded by group 3b; group 2a areas 'pass into' group 3c and so on. While different groups would define different areas

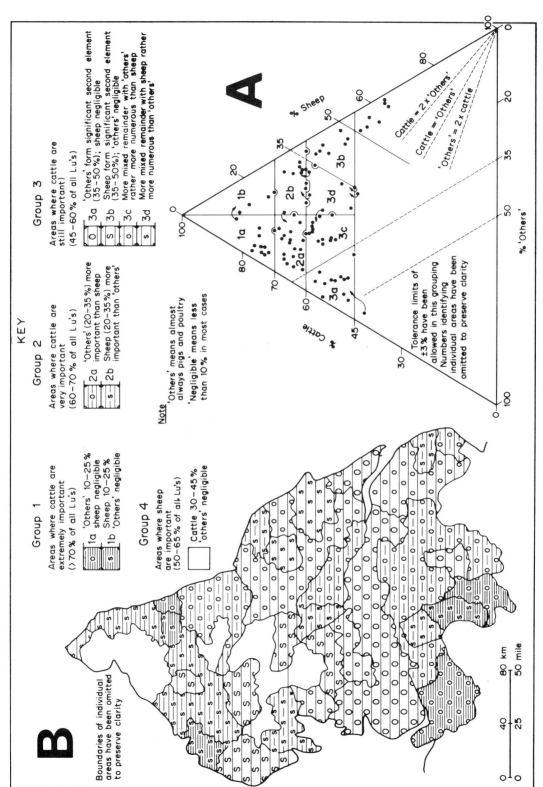

Figure 27 Characteristic combinations of livestock in agricultural areas in Northern England (1958). Figures refer to percentages of total livestock measured in *livestock units (l.u's)*, a comparative measure based on feeding requirements.

Source: Author's map from figures in J. T. Coppock, 'Agricultural Atlas of England and Wales'.

they would not destroy evidence of these trends, which are an important part of the message of the map in Figure 27B. As in Figure 26B, the shadings used to represent various groups try to emphasise and to some extent depict the characteristics and contents of each group.

B DEALING WITH MORE THAN 3 VARIABLES

When more than 3 variables have to be grouped the limitations of 'flat-sheet' and therefore two-dimensional solutions becomes obvious, even the 3 'dimensions' of the triangular graph being in reality a special case, only possible when *components*, i.e. values adding up to 100%, are involved. Yet multivariate situations are common enough; how, then, can they be mapped?

One possible solution, where all the variables *are* components, is to regroup them into three basic and logical combinations, using a triangular graph to sort out groupings of these three combinations and on the resultant map indicate the presence of the original variables by means of letters or symbols. An application of this idea to the mapping of 17 agricultural land uses in Sweden, regrouped initially into 'grass and hay', 'animal crops', 'cash and other crops', will be found fully described in Chapter 7 (pp. 155–159).

Where very large numbers of variables are present some preliminary combination of them seems inevitable but where the number is still small, say 6 or less the following development of the scatter graph may be of assistance. Though described here with yet another *componental* situation (total agricultural activity (man-days) in regions of Britain, broken down into 5 main headings)[1] it is possible that it might also be adapted to deal with say 3 or 4 unrelated variables.

Although the basic device used here is still a scatter graph two main departures of principle are involved, namely (1) the two axes *no longer relate to specific variables* but are used for the largest and second largest components whatever these may be, and (2) the *values* (not the type) of remaining components are represented by adding symbols at the plotted points. In this case a square was added to represent any other components of value 20% or more, a circle for one of value 15–19% and a dot for one of 10–14%; where more than one such component occurred the symbols were repeated preferably either 'nested' or interlinked and if no such variable occurred the position was marked with a cross.[2] The method places a good deal of reliance on two basic assumptions, namely (1) that in many areas the two largest components will tell much of the story and (2) that many of the components are trivial and can be neglected in so complex a picture, and this is borne out in practice. The original data for Figure 28 contained 1715 values—5 components for each of 343 areas— but one third of these were responsible together for only 8% of all activity and none individually exceeded 10% of the activity in any one area; on the other hand the top 280 values accounted for no less than 40% of all activity.

Further comments on the method are, perhaps, best left until an examination has been made of Figure 28B on which this data has been plotted. There, as in previous

[1] Dairying; other livestock; cash crops; pigs and poultry; horticulture.
[2] With so much information to plot a very large scatter graph is essential. The original of Figure 28B measures some 30 × 24 inches. Even so 'nesting' symbols—square, circle, dot are essential for compactness.

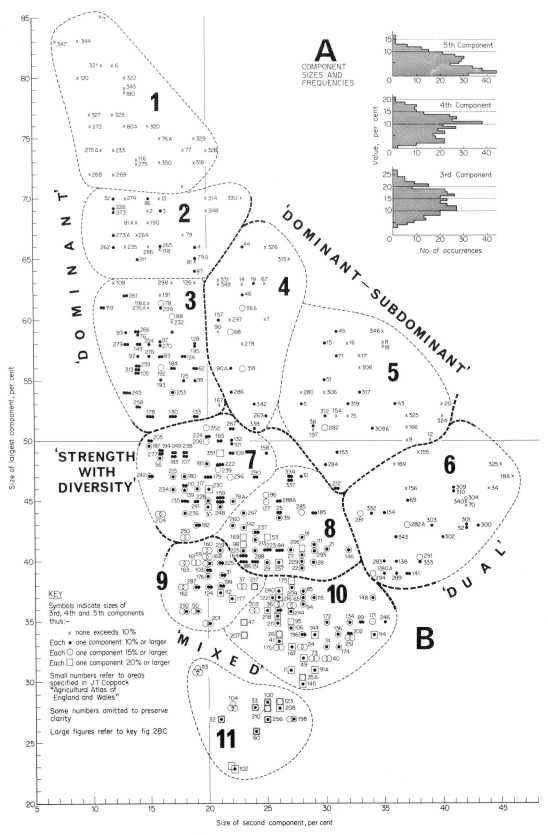

Figure 28A, B and C Agricultural Regions in England and Wales classified according to activity types (1958).

Source: Author's map from figures in J. T. Coppock 'Agricultural Areas of England and Wales'.

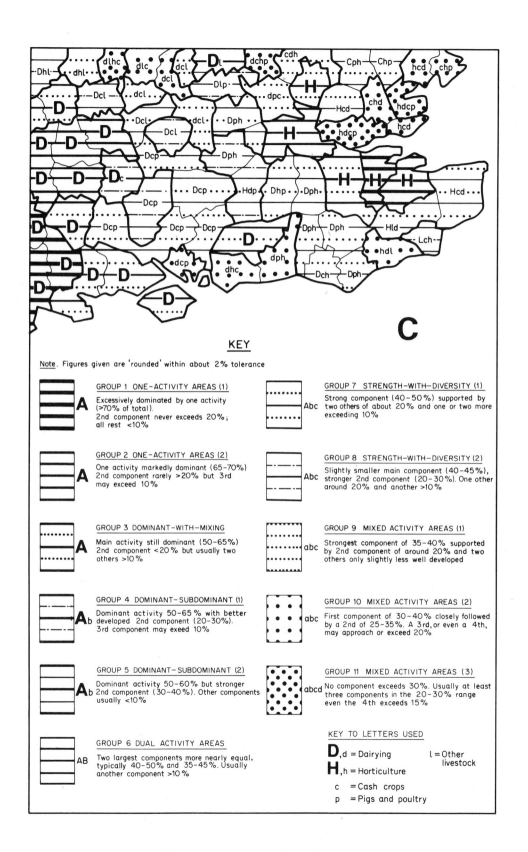

C

KEY

Note. Figures given are 'rounded' within about 2% tolerance

GROUP 1 ONE–ACTIVITY AREAS (1)   A

Excessively dominated by one activity (>70% of total).
2nd component never exceeds 20%; all rest <10%

GROUP 2 ONE–ACTIVITY AREAS (2)   A

One activity markedly dominant (65–70%)
2nd component rarely >20% but 3rd may exceed 10%

GROUP 3 DOMINANT–WITH–MIXING   A

Main activity still dominant (50–65%)
2nd component <20% but usually two others >10%

GROUP 4 DOMINANT–SUBDOMINANT (1)   Ab

Dominant activity 50–65% with better developed 2nd component (20–30%). 3rd component may exeed 10%

GROUP 5 DOMINANT–SUBDOMINANT (2)   Ab

Dominant activity 50–60% but stronger 2nd component (30–40%). Other components usually <10%

GROUP 6 DUAL ACTIVITY AREAS   AB

Two largest components more nearly equal, typically 40–50% and 35–45%. Usually another component >10%

GROUP 7 STRENGTH–WITH–DIVERSITY (1)   Abc

Strong component (40–50%) supported by two others of about 20% and one or two more exceeding 10%

GROUP 8 STRENGTH–WITH–DIVERSITY (2)   Abc

Slightly smaller main component (40–45%), stronger 2nd component (20–30%). One other around 20% and another >10%

GROUP 9 MIXED ACTIVITY AREAS (1)   abc

Strongest component of 35–40% supported by 2nd component of around 20% and two others only slightly less well developed

GROUP 10 MIXED ACTIVITY AREAS (2)   abc

First component of 30–40% closely followed by a 2nd of 25–35%. A 3rd, or even a 4th, may approach or exceed 20%

GROUP 11 MIXED ACTIVITY AREAS (3)   abcd

No component exceeds 30%. Usually at least three components in the 20–30% range even the 4th exceeds 15%

KEY TO LETTERS USED

D, d = Dairying   l = Other livestock

H, h = Horticulture

c = Cash crops

p = Pigs and poultry

methods, points have to be sorted into groups which should be similar not only in plotted position but also in their other characteristics as described by the symbols, and fortunately, as Figure 28B shows there is some consistency, crude spatial groupings tending to have fairly consistent 'symbolisation' as well. As always the grouping is not easy and is best made with a consistent theme and the ultimate description of the results borne in mind; in Figure 28B an eleven-fold grouping based on declining degrees of dominance through to duality and 'mixed' was selected.

Unlike the previous methods described in this section grouping is not the end of the process. All that grouping of this kind has achieved is a description of activity *patterns*, it says nothing about *which* activities are concerned in areas classed as 'strongly dominant', 'dual' and so on. The only way of including this information is to supplement the group shading by symbols or letters which 'spell out' its detailed implications and a portion of the resultant map of agricultural activity in England and Wales, incorporating these ideas is shown in Figure 28C. There is no disguising the fact that Figure 28C is an extremely complicated map—though what else would one expect. Complex and compound situations inevitably produce a map with a complex and compound message which can be appreciated only after detailed study, and certainly *not* at a glance. Yet there is no doubt that such appreciation of compound situations is necessary (why else do we produce the statistics?) and that, if this is so, the non-instantaneous but still appreciable message of the multivariate statistical map can convey a most valuable overall impression and description that the bare statistics themselves could never give or that, expressed descriptively, would take pages of text.

Like Figure 27B, Figure 28C is not an 'inevitable' end product; other groupings would have produced slightly different maps, though these would, as before, have indicated common trends. Even so the complexity of any of the possible grouping solutions clearly gives the lie to the alternative ideas of deriving much of real value by facile comparison of a series of maps, or of inspecting a single map studded with multitudes of pie graphs or similar symbols.

As with previous devices used in this chapter the theoretical simplicity underlying Figure 28 conceals many real practical difficulties. Critical amongst these is the choice of values for the symbols on the scatter graph (Figure 28B) and the design of shading-letter combinations which would ultimately represent the characteristics of the groups. In effect the choice of symbol values was determined by an inspection of the frequencies and magnitude of all 5 variables individually, as portrayed in Figure 28A. There the choice of 10% as the lowest value represented by a symbol can be seen conveniently to eliminate the need to show all but 37 of the 343 5th components and about half of the 4th components as well, whilst using 20% for the upper level ensured that several important 3rd components get adequate representation without 4th components being involved except in one case. The use of round numbers helps assimilation and a 15% value neatly divides the 10%–20% category into two halves. A great deal also depends on the general 'run' of the data, widespread and varying values as here[1] being much more amenable to this kind of treatment than less distinctive combinations.

[1] E.g. 86, 10, 3, 1, 0 at the dominant end and 26, 24, 22, 17, 11 at the 'mixed' end

With the shading/letter design the lack of colours imposed a severe handicap which was resolved by using a range of shading which attempted to reflect the predominant characteristics of the group—heavy shading indicating 'dominant' groups, 'dot' shading mixed groups and combinations and variants of these various intermediate groups. In a same way the number and style of letters allowed with each group was chosen *to emphasise the distinctive characteristics of that group.* It was tempting to use letter symbols of particular kinds to represent *all* and *only* components which exceeded particular values but this was rejected. Because of slight inconsistencies inevitable in the grouping the number of letters used would have varied from area to area within the same group, making recognition of groups more difficult and re-introducing some of the complexity which the grouping was intended to remove.

C MATHEMATICAL SOLUTIONS TO MULTIVARIATE PROBLEMS

In contrast to the simple and imperfect sorting and classifying techniques just described are the highly sophisticated mathematical techniques such as cluster analysis and factor analysis which are coming into increasing use to describe and analyse multivariate situations. Whilst the superiority of such techniques is evident the simpler 'hand-methods' do at least have the virtue of allowing participation in the fascinating and instructional business of assessing multivariate situations without the need for access to computing facilities which the more sophisticated methods demand, and which inevitably reduces their practicability. There is still much that can be learned from even a simple 'hand-operated' investigation into a multivariate situation, provided that the maps produced are thought of as an *aid* to understanding, a *means* to an end, rather than an end in themselves. However produced and on whatever basis there are few statistical maps which have any right to claim to be much more than that.

5 Choosing the right method and making the most of it

A review of the techniques which are available and the statistical pitfalls which may be encountered has placed us in a position to decide how to choose a technique for a particular task and begin the actual plotting of the information. The couple of dozen or so methods which were described in Chapter 2 will have made it apparent that in dealing with most statistical distributions there *is* a choice of method available and this chapter describes some of the broad considerations which may influence the final decision.

One important point which emerges from Chapter 2 and Figure 2 (p. 17) together is that the amount of choice will depend greatly on the type of statistical base which is being used, figures for point values, or values along a line offering less possible variety than figures relating to areas where a wide choice is possible. A useful rule to adopt in these cases is that the best method is the one which *looks* right and which conveys an impression which most closely resembles the kind of distribution which the cartographer envisages in his own mind.

Figures 29A, B, C and D show different ways of illustrating the distribution of population in South-west Lancashire. The figures used were the 1951 populations of civil parishes and in Figure 29A this total has been represented by proportional symbols, one symbol for each parish. In Figure 29B a dot map with one dot for every 250 persons is used whilst Figure 29C combines these methods, using dots for rural areas but proportional symbols for all units over 10,000 population which appeared to have definite urban quality. Figure 29D turns to another aspect of the figures and illustrates the distribution by means of a density shading map. If the main aim of such a map was simply to present a general picture of population distribution in south-west Lancashire, describing where people were and in what manner they were distributed on the ground, the author's preference would undoubtedly go to Figure 29C which most nearly corresponds to his impression of the way in which the population is actually distributed—a rural scatter of population varying in density with some areas quite thickly peopled (as in the coalfield for example) interspersed with important urban centres of varying size. Figure 29A is too rigid to be satisfactory, for the division between symbol and background is too clear cut to reflect reality; Figure 29B is much better and conveys a good deal of the impression needed, especially where the dots coalesce at the larger centres. Where Figure 29C scores over Figure 29B is that by introducing a new type of symbol it is able to distinguish sharply between population concentrated in the main urban centres and that in other areas whilst at the same time it offers a chance of giving positive information about the size of those centres, which Figure 29B cannot. Figure 29D, by dealing in density, introduces a

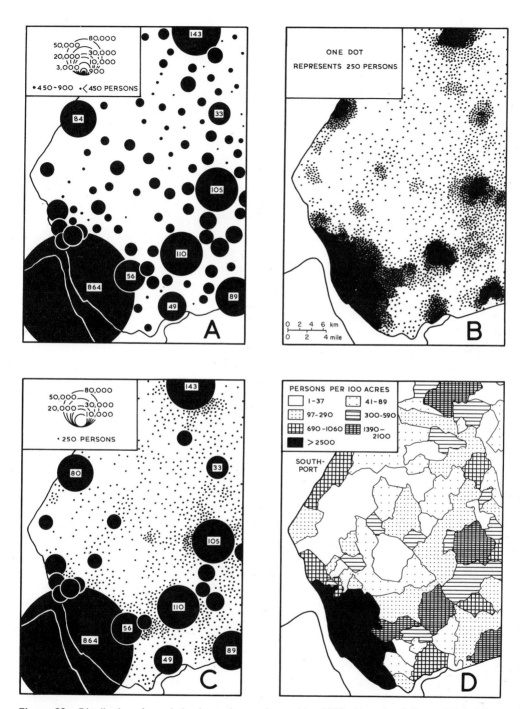

Figure 29 Distribution of population in south-west Lancashire, 1951, shown by: **A** Proportional circles for the population of each civil parish. **B** Dots, one dot for every 250 persons. **C** Proportional circles for the population of each urban centre over 10,000 population with dots for the remaining areas. **D** By shading according to the population density in each civil parish. The whole figure offers a useful opportunity to see the different impressions produced when the same information is treated by four different methods. This diagram is freely adapted from an idea in F. J. Monkhouse and H. R. Wilkinson, *Maps and Diagrams*, p. 236, London, 1952.

Source: Registrar-General, *Census of England and Wales, 1951*, County Volume, Lancashire; London (H.M.S.O.), 1954.

rather imaginary measure-of-convenience for assessing this situation. If we were interested in density *for its own sake* there would be little to complain about in Figure 29D, but if we are using it merely as an indirect way of summarising a distribution then its use seems far less satisfactory and it gives a much more incomplete picture than either Figure 29B or C.

The use of a 'joint' method for dealing with a situation which has two basic components—urban and non-urban population—seems a sensible way of fitting a technique to the actual figures. It is not without its difficulties of course; the larger circular symbols sometimes overlap their area boundaries and make difficult the dispersal of dots within their neighbours, and where this occurs the dots are either omitted or, better, used to dot in the unfortunate 'blanks' which would have occurred at those points where the same circular symbol did not in fact reach the boundary line of the area which it represents. The small statistical 'fiddling' which this involves is not likely to upset the basic picture on the map, and it would have been foolish to abandon the whole method, with its useful overall impression, for the sake of one small point of difficulty.

Statistics relating to values within a given area offer the widest choice of possible techniques for their illustration and, as might be expected, it is with figures of this kind that there is the greatest danger of choosing a method which will look wrong. We have already seen that when shading and isoline maps are being used figures must be expressed in a way which will not be unduly influenced by variations in area, but even with quantitative techniques, such as dots evenly distributed throughout the area, it is easy to allow the effect of area to come through too sharply. No difficulty arises if the quantity being mapped is, in fact, tied to land area, for example acreage under a certain land use or even simple population distribution; but if its basic distribution is remote from consideration of area then a wrong effect may ensue.

Figure 30 shows an illustration of this point. Figure 30A represents the distribution of home areas of a student population by a dot technique. Information on this distribution was only available easily in terms of geographical counties and since further breakdown of the figures would have needed much more lengthy research the dots have been distributed evenly throughout each county with the added convenience of a summary total in the middle of each county area, any dots which would have been covered by this circle being omitted. The effect is not noticeably bad in the most extreme cases such as Lancashire and the West Riding where very high numbers still produce a dominant impression, but difficulties begin with the middle range of numbers where a visual effect is obtained which is by no means satisfactorily related to quantity. Contrast for example the effect produced in County Durham, the North and East Ridings by figures not particularly dissimilar in size. Here the density of dots and hence the visual impression is markedly affected by the *size* of the county concerned so that the larger North Riding comes out with a relatively much weaker impression. This is not at all the effect that is needed, for there is little or no relationship between what might be termed 'student-supply-potential' and superficial area and it is not desirable, therefore, to have this aspect of area influencing the map. Student-supply-potential, if it is related to anything at all, will be more closely connected with population than anything else and since population in Britain has no definite relationship to superficial area the method has failed.

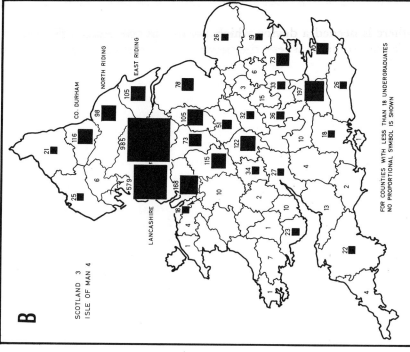

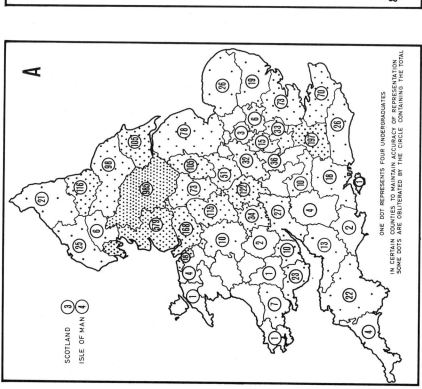

Figure 30 University of Leeds—home areas of undergraduate students, 1958–9. Of the two methods used the first is the least satisfactory, since it allows the impression of quantity to be affected by the area of the county, a factor which is quite irrelevant to the theme of the map: the second method avoids this pitfall.

Source: Information supplied privately by the University authorities.

To replace it there is needed a device which is not in the least affected by variation in area, and this is one situation where the single proportional symbol, as used in Figure 30B, can successfully deal with statistics relating to areas.

In the example just described the dot symbol which *seems* so suited to representing a scattered group of people turns out to have unsuspected limitations and indeed, as Chapter 2 indicated, most methods in the group relating to areas have their own idiosyncrasies of effect. Dot maps are weak in quantitative impression, despite a quantitative base, excellent for showing general distribution, but poorly adapted to illustrating *small* differences should these be important. All too often one sees dot maps which, despite the greatest care and meticulous preparation, can be said to give no more than a general impression of fairly even distribution over the mass of the map with only a few exceptional areas standing out by coalescence of the dots—a great deal of work for a very general result. Shading maps, on the other hand, are excellent for emphasising differences (even quite small ones if necessary), give a clear, positive statement of quantity but may give a poor and rather artificial picture of distribution (e.g. Figure 29D); isoline maps are best suited for illustrating distributions which show regular trends but in other ways they have some of the defects of shading maps. Density of colour is yet another factor of the areal techniques which needs matching to the actual figures and consideration of this aspect has already been made in Chapter 2.

A single example will serve to summarise many of these points. Suppose we have within an area a relatively even distribution which nevertheless shows small but important variations. A dot map would be of little use in representing such a distribution; a shading map with carefully planned shading limits would be better or, if the variations appeared to have a rational trend, even an isoline technique could be used. In either of the last two methods, however, range of colour would have to be kept fairly small so that once the method had picked out the differences they were not further emphasised out of proportion to their actual size.

Once the important choice of a particular method has been made it is obviously important to use it to its best advantage. Even the most well-selected techniques will not be put into practice without certain difficulties being encountered and the successful surmounting of these can add considerably to the finished effect of the map.

One of the most recurrent of these difficulties arises from the wide range of values which so often have to be accommodated. Chapter 2 showed that increased range may sometimes be dealt with by technical means, e.g. the use of areal or even volume symbols, but even so these devices frequently leave problems so far as the most extreme values are concerned. In many cases no completely satisfactory solution will be possible and the cartographer should guard against the danger of spoiling the design over 90% of the map simply by attempting to accommodate a few extreme values. In particular this applies to finding suitable sizes for proportional symbols which represent very small values in an extended range. Successful treatment of these lower values may well result in the bulk of the symbols being over-large and unwieldy and rather than do this it is worth while deciding whether these values are important enough to bother showing at all or, if so, whether all occurrences below a certain value cannot be shown by a uniform and very small symbol.

Extreme values at the upper end of the scale are not so easily tackled for, unlike the smaller ones, they are of extreme importance and cannot be ignored. One possible approach is to regard these values as extreme enough to amount to something quite different from the items forming the bulk of the map and if this is the case an easy choice is to 'break' the method being used and adopt an entirely different way of representing them. A very commonly recurring example of this is encountered in mapping rural populations by the dot technique. A dot value which is adjusted to small villages of 200 or even less population will not easily show a market town of, say, 20,000 population and even less an important local centre with 100,000. The village and the sizeable town are too unequal and dissimilar to fit into the same technique easily and, as Figure 29C showed, a combination of two methods each capable of treating one of the two components will be much more successful.

The rule never to spoil the whole map for the sake of a few extremes can be applied in other contexts. The areas into which symbols have to be fitted are rarely of a suitably convenient shape and there will almost always be one or two areas somewhere which are extremely small but which contain a large quantity and therefore need a sizeable symbol. Once again the most satisfactory solution is to use the symbol sizes which the rest of the map dictates and accept overlapping as inevitable. When Figure 29C was being worked out the unusual elongated form of the County Borough of Southport presented a problem of this kind. As an important centre of 84,000 people a circular symbol was needed, but this would have left two 'tails' at either side completely empty. Unfortunately, the circle for Southport did not displace elsewhere any other dots which might have filled these areas and to avoid the blanks on the map the method was modified slightly by cutting down the size of the symbol to 80,000 and distributing the last 4,000 in the form of 16 dots over the remainder of the area.

Devising a scale of symbol-sizes to suit a wide variety of values is not an easy task and an effective method of testing any suggested scale can be of considerable assistance. One such test is to try out the sizes envisaged on representatives of large, medium and small values and on values occurring in large, medium and small areas, avoiding extreme cases. If a reasonable fit is obtained in these six cases the scale will probably not give rise to serious difficulties, although unusual combinations of value and area may still present the odd problem.

Yet another example of the occurrence of 'awkward' values may be revealed when a dispersal graph has been drawn to determine the grouping of values for a shading map. It may be that the graph reveals well-established and narrowly defined groups generally, but with one or two values lying rather isolated between groups. The usefulness of the group in categorising sizes might be considerably weakened if it had to be extended very considerably to accommodate these isolated cases and there seems to be no reason why the group limits cannot be defined to exclude these exceptional areas; these could still be shaded in accordance with the most similar group, but marked conspicuously with an additional symbol, such as an asterisk, to indicate 'Exceptional value—not falling within the given limits of the group but so coloured because it most nearly approximates to this type'.

It will be apparent by now that the rendering of statistics into pictorial form is far from being a simple procedure, involving as it does fundamental decisions on techniques and the surmounting of difficulties inherent in the statistical material itself. More than this it involves a good deal of tedium also, for there may be long calculations needed before the method can begin and the actual plotting of the information on paper is usually a laborious and meticulous job. With as many and varied difficulties as these to be overcome it is easy to believe that they are the whole problem and there is nothing more to be considered. Unfortunately, this is not so. The *successful* presentation of a series of statistics does not end when they are committed to paper; there is a final link in the chain yet to be made—namely, the satisfactory use of the map by an observer. Few aspects of statistical presentation are more neglected than this, so much so that one feels bound to conclude that in many cases the possible difficulties of the map user have scarcely entered the draughtsman's mind and this important question of *legibility*, which has been touched upon earlier in this book, clearly demands rather more consideration here.

It is most important that when statistics are presented cartographically the resultant map should be capable of giving back to its user after a minimum of inspection and effort a clear indication of the information it was designed to carry and the main points it has to make. In the vast majority of cases this message, and the portrayal of quantities on the map, is made through the medium of abstract symbols, such as circles, bars, types of shading or a scatter of dots, which will often have been designed in such a way that they bear a distinct proportional relationship to the quantities represented. The ability of the map user to judge the significance of the size or arrangement of such symbols is therefore an important link in the chain of transferring messages from the map to its user; and the continued use of these geometrical devices seems to presuppose that many cartographers are quite satisfied that the average map user is possessed of sufficient ability to enable him to appreciate their implications. With this opinion the author disagrees. He has long sustained considerable doubts concerning ordinary people's abilities in this field and to test the validity of this belief conducted on students in the Department of Geography at the University of Leeds, an experiment which took the form outlined below.

Seven common methods of representing quantitative statistics were tested—namely, proportional bars, squares, circles and spheres, pie graphs, dots and proportional shading. For each of these methods a large card 60 cm × 45 cm was prepared on which were displayed ten values (hereafter referred to as 'quantities') and the cards were successively exhibited for a period of about four minutes to the students who were told the value of the smallest quantity on the card and asked to estimate the remaining nine. Exposure of each card was prefaced by a very brief explanation of the method involved. A second exercise was derived from the same cards by presenting the values of the smallest and largest quantities and estimating the intervening eight. To provide a different set of values from those of the first exercise the numbers involved were multiplied by a constant and, as over half an hour elapsed between the first exercise and the second, there seems little reason to believe that the second exercise was influenced by any memory of estimations made during the first. A reproduction of the cards and values used is given in Figure 31. The experiment was tried with two separate classes, one of 20 second-year students studying for General Degrees, the other of 24 first-year students studying for an Honours Degree in Geography; the time allowances which had proved necessary for the first class were automatically imposed upon the second.

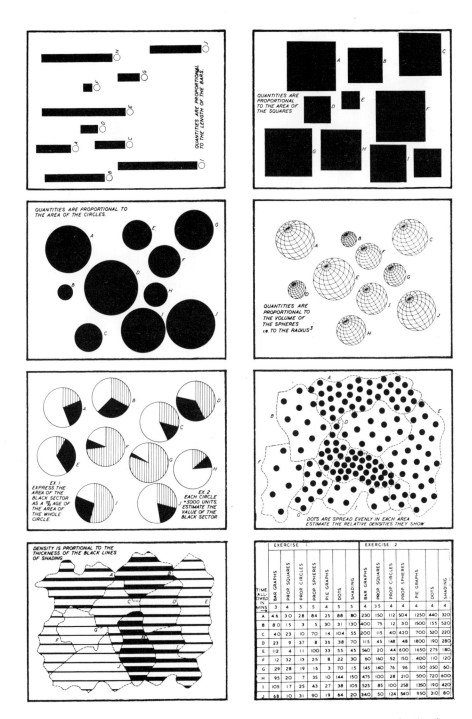

Figure 31 Reproductions of the cards used in the experiment to test powers of estimation. The actual values represented are given in the bottom right-hand rectangle.

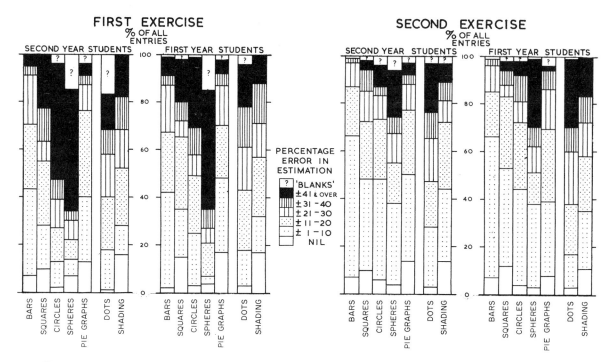

Figure 32   The results of the experiment, showing the percentage of errors of different sizes obtained from each method.

The results obtained from the exercise (shown diagrammatically in Figure 32) are most interesting, particularly in the close similarity which occurs between the answers from the two quite separate classes. All estimates for the quantities in each method have been classified as either correct or falling into one of five classes of error, the error being expressed as a percentage of the true value of the quantity which was being estimated. The percentage of the total estimates for each method which fall into the various classes of error are differently shaded and can be read on a central vertical scale. Blanks, caused by incomplete entries, are also recorded.

The most outstanding feature in the results of the first exercise is the magnitude of the errors involved. Errors of ±20% and more are quite common, applying to never less than 24% of the estimates made (in pie graphs) and rising to as much as almost 80% of the total (in spheres). Quite a large proportion of the entries were in error by values ranging from ±40% to 100% or more. The best results in the experiment are noticeably those relating to linear estimation, i.e. bars, proportional shading and pie graphs[1] where half or more of the estimates were within ±20% of the correct answer. Proportional squares and circles proved more difficult of estimation; circles generally produced the less accurate results, although it should be noted that variations between the two classes in the accuracy of the estimation of circles in the first exercise provides the only instance of marked difference in the order of magnitude of the results obtained from the two

[1] Pie graphs involve linear estimation if the length of the arc bounding the sector is considered.

separate sources. Proportional spheres proved quite hopeless, the really large errors involved[1] indicating that the implications of the method were far beyond the comprehension of many students; on the other hand the results from dots, where relative 'blackness' had to be estimated proved much better than expected, about two-fifths of the estimates falling within $\pm 20\%$ of the correct value.

The second exercise showed that in the new conditions of estimation the *relative* accuracies of the various methods remained very little changed, but individual estimates were markedly better. Nevertheless, a sizeable proportion of the estimates was still grossly inaccurate and it is noteworthy than in most methods about a third of the estimates err by more than $\pm 20\%$ and the results for dots and pie graphs were scarcely improved at all.

It may be claimed that these inaccuracies were only to be expected. We are not accustomed to making fine estimates of length or size and certainly not to thinking in terms of the squares (squares and circles) and cubes (spheres) of numbers. If we do not know the cube of 6 or the difference between 7^2 and 13^2 we are hardly likely to recognise it very clearly when it is presented to us graphically, and it is precisely for this reason, of course, that graduated scales or 'sample' scales are added to statistical maps where these methods are in use. The absence of such a scale on the cards used in the experiment was one of the important differences between the experimental conditions and normal usage of a statistical map, but unfortunately the detailed results of the exercise also cast some doubts on the usefulness of these scales as aids to *visual* estimation. An outstanding feature of the estimates was that in more than half of the results the quantities were not even placed in their *correct order* of magnitude. Difficulty appeared to be encountered wherever differentiation was required between quantities which differed by 10% or less *and which were widely separated*. Similar cases where the quantities were adjacent caused little difficulty, but this does not hold out much hope that accurate estimates will be made by visual comparison against a scale which may often be at the other end of a sizeable map. Naturally, comparison by measurement could be made, but if many values had to be so treated the general efficiency of the map could only be considered as very small.

The nature of the experiment does not render it entirely immune from criticism if it is to be used as a basis for drawing conclusions with far-reaching implications. The number of students tested was small and the processes involved were not strictly those encountered in using a well-drawn statistical map with a good scale for comparison. On the other hand it is unlikely that the *average* user will spend almost five minutes perusing one statistical map (and one which contained no more than ten quantities at that) in an attempt to analyse its quantitative relationships alone, and this tends to offset the more difficult process of unaided estimation which was required. The results of the experiment on a general level cannot be gainsaid. Here were 44 persons (and there is no reason to believe they were abnormal in this respect) who experienced very real difficulty in estimating and differentiating quantities, two processes which they are frequently required to do when using statistical maps.

---

[1] The worst was more than 900% in error and there were many more estimates well over 200% out.

This result confirms the impression that there exists a very genuine dilemma for the map designer, whose task is not merely to translate statistics into a cartographic technique but to translate them into the mind of the user *via* an easily legible cartographic technique. There arises in this connection the unfortunate paradox that the mathematical accuracy and tedious preparation of the cartographer's portrayal is matched against the comparatively inaccurate estimates and brief perusal of the average map user.

It would not be stretching the point too far to say that, on balance, a clear indication of quantity is therefore more essential than *strict* mathematical accuracy, and fortunately this clear indication is not difficult to achieve. A single small figure indicating the actual value written unobtrusively against a proportional symbol suffices at once to augment its usefulness (see for example Figure 29C (p. 103), and other methods can be redrawn in such a way that they become more legible. Scales can be added easily to devices such as proportional bars (see Figure 13B (p. 41)), and Figure 33 shows the same idea adapted to pie graphs and a proportional shading technique which has been redrawn, solid bands of shading being replaced by a pattern composed of lines and dots so that values may be more easily read off.

It is rather unusual, when one considers that statistical maps are designed to convey quantities to the eye, that numbers themselves have met with little use as proportional symbols. In this respect their more complex shapes, with consequent difficulties in drawing, have probably told against them, but there seems no other reason why they could not sometimes replace with great advantage the simpler geometrical symbols such as the square and circle which are so frequently employed. Figure 34 illustrates a successful example. The figures in the lower diagram fit exactly into the squares drawn in the upper one so that the effect of distribution and of relative values remains completely unaltered whilst the lower drawing has the inestimable benefit of presenting a clear and unequivocal statement of fact without any necessity for reference to a scale line. Where statistics relating to quantity within a given area are being plotted there is the additional advantage that the figure symbol with no rigid perimeter seems to belong to the whole area; the geometrical symbol hints at concentration in one point, which is frequently not the case.

Of course it is true that all these adaptations of common techniques involve more work for the cartographer, but this is after all his job and there is at least the certainty that his efforts will be rewarded by increased legibility of the map. The improved techniques are not always as tedious as might be imagined. The numbers in Figure 34 were easily fitted into their squares by using a rectangular type of figure, the thicknesses of the lines being automatically obtained from a pair of proportional dividers set at about 5:1; a scale of chords helps in drawing a circumferential scale for pie graphs (see Figure 30) and this scale, when completed, may obviate the need for angular calculations as well. In Figure 34 even representation by seven or eight sizes of 'UNO' stencilled numbers would produce an impression as useful as that obtained from squares or circles.

An interesting by-product of these more vivid methods is that continued use of the maps will often result in quite involuntary memorising of some of the more important figures found there. Indeed it may very well be true that the effect of repeatedly seeing the relevant information, even when portrayed in an entirely non-proportional manner as in Figure 35 which is based on a page in a children's school atlas, will ultimately produce a

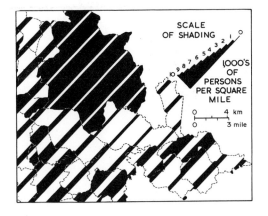
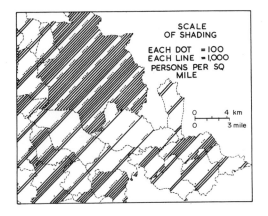

POPULATION   DENSITY   IN   URBAN   LOCAL   GOVERNMENT   AREAS   IN   THE   CENTRAL   WEST   RIDING , 1951

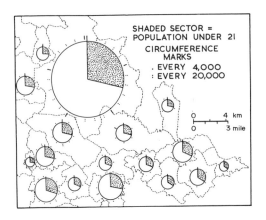
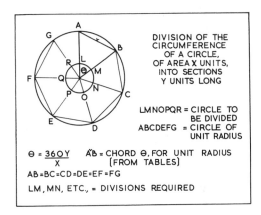

POPULATION   COMPOSITION   IN   URBAN   LOCAL   GOVERNMENT   AREAS   IN   THE   CENTRAL   WEST   RIDING , 1951

Figure 33   Two methods rendered more legible by means of altered presentation. In the upper portion of the figure solid shading (*left*) is replaced by lines and dots (*right*) which are easily counted to obtain the required density. In the lower portion pie graphs are improved by adding a scale of *quantity* around the circumference of each circle. The lower right-hand diagram illustrates a simple way of adding such a scale using a table of chords.

Source: Registrar-General, *Census of England and Wales, 1951*, County Volume, Yorkshire (West Riding); London (H.M.S.O.), 1954.

far more real appreciation of the figures and their significance than would the same information represented by proportional geometrical symbols alone. By all means let us have proportional symbols to thrust home to the user the dominant themes on any map or diagram, but let us be sure that if it is at all possible there is equally accessible information on actual quantities as well.

Once this theme is established it is surprising how widely its ramifications can proceed.

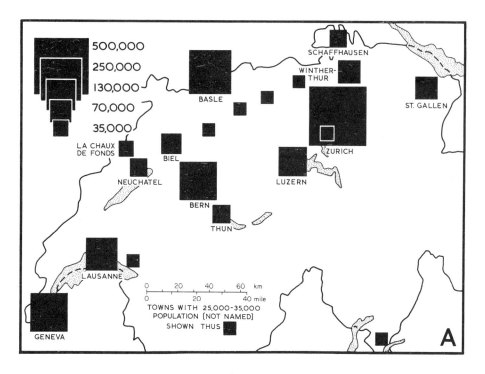

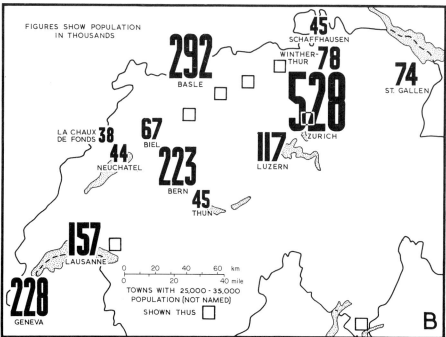

Figure 34 Switzerland—populations of the main urban centres, 1958. The two maps contrast the informative nature of a figure symbol **(B)** with the less helpful geometrical shapes (as in **A**).

Source: Eidgenossischen Statistischen Amt, *Statistiches Jahrbuch der Schweiz, 1959/60, Basle*, 1960.

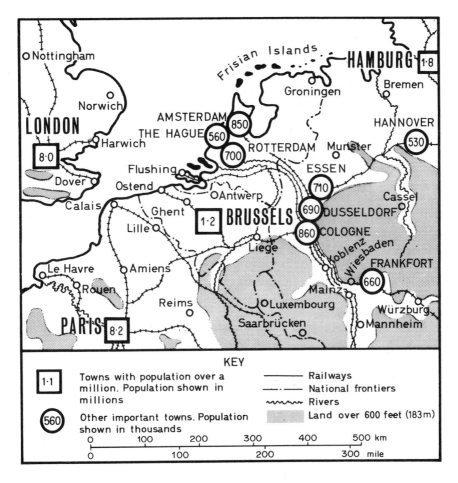

KEY

1·1	Towns with population over a million. Population shown in millions	———	Railways
560	Other important towns. Population shown in thousands	—·—·—	National frontiers
		～～～	Rivers
		▨	Land over 600 feet (183m)

0   100   200   300   400   500 km

0   100   200   300 mile

Figure 35 Part of north-west Europe, showing the populations of the principal urban centres. It seems probable that continued usage of maps with simple informative symbols of the type shown here will result in quicker and fuller appreciation of the relative size of the places shown than would the more frequently employed proportional geometrical symbols.

Source: This figure is a slightly modified version of a page in the *All Essentials School Atlas*, published by W. and A. K. Johnston and G. W. Bacon, Ltd., Edinburgh.

Figure 36 A form of layer tinting for relief maps. The incorporation of numerical designs into the shading obviates the need to refer to a key—a decided advantage where quick map-reading is essential.

Source: A.M.S. Bulletins, No. 11, September 1944.

Figure 36 shows an unusual adaptation of this idea applied to the problem of layer tinting to depict relief, the 'shading' density being derived from a printed indication of altitude seen against various backgrounds; and in Figure 37 which illustrates a Turkish stamp issued to attract attention to a national census the message is conveyed by the figures, suitably drawn, far more vividly than it could have been by any other method. In this case the simple and direct appeal of figures instead of abstract symbols was particularly apt where a very wide public had to be reached.

Legibility of statistical maps has, of course, other facets besides the appreciation of quantitative data, but since these may generally be considered as part of the wider subject of general map design, as opposed to map techniques, they are reserved for the next chapter.

Figure 37   A Turkish postage stamp issued to publicise an impending census. The use of figures, rather than symbols or graphs, brings home the message of increasing population more forcefully to a wide public.

6 General aspects of map design

With the possible exception of those which are drawn by a computer (as described in Chapter 8) every really good statistical map is to some extent an artistic creation. In common with many other devices which are used to put over or 'sell' certain ideas to a wide public, a statistical map needs 'eye-appeal' to stimulate interest and this will derive as much from good use of colour, sensible arrangement and design of components, and neat draughtsmanship, as from careful attention to statistical material and available techniques. There are very few designs for the presentation of statistics which cannot be improved by attention to the 'artistic' elements of their composition, and the purpose of this chapter is to give a brief guide to some of the more general and easily appreciated rules which govern this aspect of the work.

Naturally, in a book of this kind it is impossible to give a prolonged exposition either of the principles of artistic layout and design or of the techniques of draughtsmanship;[1] it is quite feasible, however, to introduce the statistical cartographer to the most easily assimilated aspects of both, perhaps stimulating his interest sufficiently to pursue the matter elsewhere or encouraging him to develop the habit of observing these aspects in the many maps and diagrams which he will encounter from day to day. There is no better teacher than critical observation of the work of others, and the habit of making a mental note of a particularly attractive colour combination, layout or general design offers one of the quickest means of acquiring experience and judgement in these matters. The 'artistic' elements of design can be divided into two rather dissimilar sections—namely, general layout and design and the actual drawing of the map; in this chapter the two sections will be treated separately and in that order.

The first of these is no stranger here, for observations which could sensibly be included under the heading of general layout and design have been made from time to time. We have already met the unfortunate malady of 'statistical indigestion', for example, and its counterpart statistical congestion can be equally troublesome. The temptation to overload statistical maps is understandably very great. For various reasons, such as cost or space, they may have to be limited in number so that it appears necessary for one map to serve more than one basic purpose. At times like these it is worth while remembering that *the simpler the map the clearer it will be*, and that two simple and separate maps may each show clearly what would be only a confused jumble if combined into one. Furthermore, it is not always necessary for this splitting of the presentation to occupy more

---

[1] A very useful description of these, suited to both amateur and professional alike, will be found in A. G. Hodgkiss, *Maps for Books and Theses*, (David & Charles), 1970.

space than a combined map as simpler distributions may be shown at a smaller scale than would be necessary to maintain clarity for more complicated patterns. A good example of a type of 'combined' map which frequently fails in its object is the dot map on which two *intermixed* distributions are shown, using dots of different colour or shape. Plotted separately, each distribution would give the observer a clear picture; but on a combined map this degenerates into a vague impression that the two distributions are thoroughly intermixed. Even if the original aim was to show interrelationships between these distributions it is open to speculation whether this could not be achieved more successfully by *mental* fusion of a clear impression of the essentials of two separate distributions, rather than relying on imperfect appreciation of *visual* fusion of the two on the same map.

It is not impossible, however, to make one map serve two purposes; indeed it may sometimes be a positive advantage to do so. Where statistics relate to areas, for example, two aspects of the same situation can usually be accommodated on one map, saving space and work and giving convenient juxtaposition of complementary aspects of the situation. Figure 38 illustrates an example of this and it will be seen that the combination is one of a *quantitative* aspect of the distribution which can be shown by a single symbol for each area, plus a more *relative* aspect where figures can be expressed as a percentage, density, ratio or in some other form suitable for portrayal by means of a shading technique. Point symbol and shaded background rarely interfere with one another and both distributions can be viewed separately or an attempt can be made to appreciate the combined implications of the two.

The idea of *attempting* to appreciate the combined implication is used deliberately in the last paragraph, for yet another fundamental rule of statistical cartography is the recognition of the observer's limitations in this respect. Too much of the deduction of inter-relationships or change should never be left to the map user to perform and if such correlations or changes are really important they will always merit separate representation, rather than being entrusted to chance recognition. Occasionally this can be accomplished by modifying the design of the maps, e.g. where a series of maps shows a city's physical growth the additional growth in each inter-map period could be differentially coloured, but where changes are more complex, such as variations in production at a number of units or changes in population distribution within a region, the only really safe course is to map this change separately.

Too much unnecessary background detail is yet another factor which may confuse or obscure the message of a map. Unnecessary background information will inevitably detract attention from the important main distributions, and features such as rivers, towns, names and boundaries not directly related to the work in hand should be eliminated or kept to a minimum. This minimum may, however, be made to serve a very useful purpose by acting as a 'geographical framework' to the main subject of the map, forming a guide to position and offering useful 'pegs' on to which may be hung less familiar aspects of the distribution. In this connection it is always wise to ensure that where places are referred to in any text which is illustrated by maps, those places are also *marked* on the maps. Few features can be more irritating than finding that full appreciation of a text is rendered impossible because additional atlas maps are not to

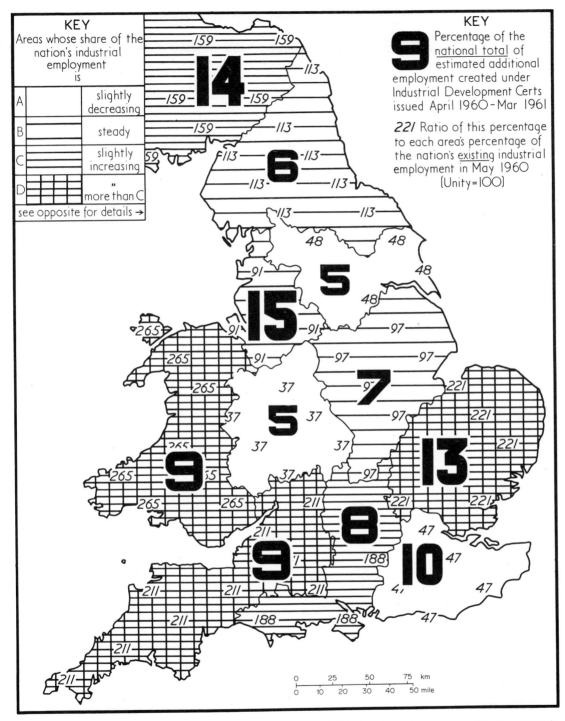

KEY

Areas whose share of the nation's industrial employment is

A		slightly decreasing
B		steady
C		slightly increasing
D		" more than C

see opposite for details →

KEY

9 Percentage of the <u>national total</u> of estimated additional employment created under Industrial Development Certs issued April 1960 – Mar 1961

221 Ratio of this percentage to each area's percentage of the nation's <u>existing</u> industrial employment in May 1960 (Unity=100)

Figure 38 Two aspects of industrial development in Britain, 1960–61, shown on the same map. The two methods employed—quantitative symbol and shaded background—do not conflict with one another.

Source: Local Employment Act, 1960: First Annual Report by the Board of Trade, (London (H.M.S.O.), 1961.

hand. Any place so marked can be regarded as of special interest and there is no need to add all places of equivalent status merely for the sake of consistency.

Attention to points of detail will often improve a map as much as will observation of the broad principles of design just described. For example, various types of shading or symbol may be designed so that they have widely differing visual appeal and dominance and this potential variation between 'dominant' and 'less significant' devices may often be used to give differential emphasis to points made on a map, or to clarify a confused situation. Figure 39 illustrates this aspect of several designs for shading and symbols. Particularly outstanding are solid colour (especially when used as the final stage in a range of shading), bold stripes and chequerboards so that areas treated in this way can readily be expected to dominate the finished map. Ideas such as this have been freely employed in designing the illustrations for this book. Thus in Figure 10 (p. 33) an important contrast had to be made between the concentration on short-distance sites in the conurbation towns and longer distance solutions for Inner London and Birmingham; the use of solid colour and chequerboard for these two elements and the 'playing down'

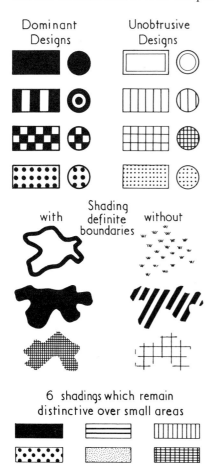

Figure 39   Aspects of symbol and shading design.

of all other varieties with less clamorous shading ensured sufficient emphasis being given to this aspect of the figures. Similarly in Figure 17 (p. 63) bold stripes were deliberately allocated to the shading for grassland, which forms so important a proportion of land use in the Netherlands.

Shading, whether applied to areas or as a filling for symbols frequently forms so important a visual element in the appearance of maps that it is important to use designs which are both harmonious and distinctive. Certain observations on this point were made in Chapter 2, but it will not come amiss to elaborate them here. Distinctiveness is best produced by ensuring variation in *density* not simply in design, little distinction being achieved, for example, between line shadings which vary only in the *direction* of the lines. Several line shadings used together may also prove rather irritating to the eye unless there is considerable complimentary variation in density; a series of dot stipples, though difficult to produce by hand, will sometimes provide clearer contrasts and more harmonious contacts at boundary lines than will line shadings, especially if the latter are of rather an open type.

Certain circumstances will occasionally prompt the use of a shading or stipple of a particular type. Broadly speaking, shadings and stipples are either compact enough to have a definite boundary or so open as to have only vaguely defined limits (see Figure 39), and this second type will be found particularly useful for portraying a distribution where exact boundaries are unknown. One of the most fundamental drawbacks of cartographic, as opposed to literary, description of a distribution is that the map is so much more unequivocal, finding it difficult to express transitional or ill-defined areas which may be suggested by the statistics and which could easily be described in words. Anyone who has tried to plot the exact distribution of marshy areas will appreciate how very well adapted to its purpose is the open scattering of 'tufted grass' symbols which are normally employed on topographic maps; the scatter of symbols has no absolute boundary and the cartographer is spared the embarrassment of delineating a most indefinite boundary line. Similar reasons prompted the use of open symbols for depicting the major regions shown on Figure 12 (p. 38).

Sound principles of design should, of course, be applied to symbols as well as to shading. Where space permits it is convenient for symbols to have some representational element in their design so that reference to a key is kept to a minimum; failing this other 'logical' designs can be introduced. Thus all symbols referring to related activities might incorporate the same basic shape as in Figure 12 where all metal industries have square symbols, wood industries round ones, mining triangular and so on. Figure 40 shows another attempt to improve the legibility of a sketch map simply by altering the design and relative emphasis of the devices used. Particular prominence has been given to those elements which were the main concern of the map—namely, proposed modernisation and improvements, and a uniformity has been imposed upon features of all kinds which are to disappear by the incorporation of an obliterating cross into their particular symbols. The wider range of overall tone and the clearer emphasis of the main features make for a more varied and interesting map.

Sensible arrangement of components is yet another aspect of symbol and map design which has already been touched upon in Chapter 2, particularly in connection with

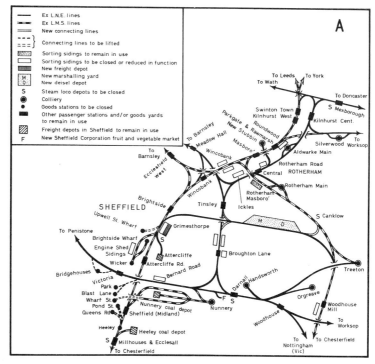

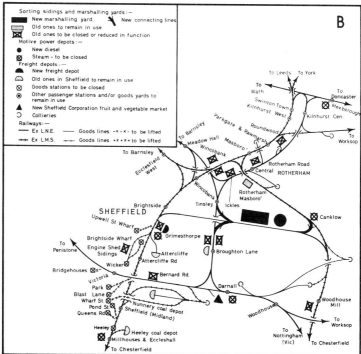

Figure 40 Railway freight modernisation schemes in the Sheffield area. The two maps show the same information but use differently designed symbols. In **A** these transfer attention mainly to layout, existing stations, etc., but in **B** (*opposite*) they focus attention on to the modernisation schemes, the real subject of the map.

divided rectangles and compound graphs. Wherever several components are involved it will usually be found worth while to devote some thought to their detailed arrangement, e.g. in Figure 17 (p. 63), where pie graphs are used, the two most important elements—arable and grassland—have been placed on either side of '12 o'clock', these being the two sectors whose position in the circle is the most stable and which therefore offer easiest comparisons of important constituents over the face of the map.

If a moment's attention paid to detailed points of this sort will rarely transform a bad map into a good one, it will nevertheless offer slight but significant improvements in appearance; and the advantages of design are all the more welcome because good designs rarely demand any more time and effort than bad ones.

Interpretation of the devices used on a map is achieved through a full key or legend which must be considered an integral portion of the map. Sufficient space to accommodate this key must be allowed when the overall design of the map is being worked out, and it is often surprising how much room a full key will in fact occupy. It is, presumably, this lack of room which is responsible for the keys one occasionally meets which contain no explanation but merely a series of numbers which are themselves explained in the titling below the map. The imposition of a 'key to a key' merely provides a further barrier between the map user and the realisation of the message of the map and its use is to be deplored, particularly because it would have been unnecessary if the map had been properly designed from the beginning. In addition to the key each map should incorporate a scale of distance, and many cartographers are tempted to add a north point as well. This latter device is often more decorative than necessary and need not be regarded as essential unless the map departs from conventional usage and has north in some direction other than at the top of the map. A suitable title will complete the essential ancillary equipment of any map, but it should be noted that where maps are being drawn to accompany a printed text it is usual today to omit the title from the design of the map and instead to set it up in type underneath the map, along with any necessary annotation; the illustrations used in this book have all been designed in this way.

Even the best designed statistical map, incorporating a sensible basic technique and attending to the detailed points described in this chapter, will not automatically have eye-appeal, and it is important to consider how this essential aspect of visual attraction may be incorporated into the design. From an aesthetic point of view, general attention to the layout of the elements which make up the map will help. The relative positions of title, scale, key and the main body and margins of the map can sometimes be varied so that a more 'balanced' composition is achieved, but all too often there is little opportunity for variation, basic shapes being such that these devices have to be restricted to those parts of the map which are not otherwise needed. In practice a much more widely applicable method of contributing eye-appeal will be found in sensible use of colour.

Few people are able to restrain their curiosity when their eye chances on some attractive, vividly coloured map, and though upon examination the subject matter may prove insufficient to retain their attention the fact remains that initially this was not a prime consideration and their attention *was* attracted. Few cartographers could ask for more than this chance of catching the eye of the public and such a potential in the use of colour obviously demands exploitation to the full. The many brightly coloured maps which one often encounters today indicate a widespread awareness of this amongst modern map-publishers and provide a welcome change from the weakly uninteresting cream, pale pinks and pale blues so often encountered in 'Commercial Atlases' of thirty years or so ago.

Whilst it would not be practicable here to enter into a prolonged consideration of the principles of colour harmony or the means of determining striking and pleasing colour combinations, a few very elementary rules can be described. In general it will be found advisable not to use too many different colours on any one map; the greater the number of colours the more difficult will it be to ensure that all of them exist harmoniously side by side, as they will almost certainly have to be somewhere on the face of the map. Similarly, in maps for which the bulk of the area needs colour in fairly large patches the juxtaposition of too many bright colours should be avoided, much better results being obtained from the use of only one or two bright colours with slightly muted tones for the rest of the areas.

Another common type of map is one where considerable areas of background colour are needed, against which small but important points of detail, such as symbols, have to be shown. Where this occurs it is best to have for the background a very 'tolerant' colour, such as pale yellow or pale grey, which need not be a 'flat' grey but which can incline towards blue, green or brown, for example, to harmonise with other colours used. Both of these backgrounds will carry almost all other colours without disharmony and the detail can therefore be picked out in very bright, clear colours such as scarlet, cobalt blue, medium green or of course black. Greens, blues and browns can also be used as background colours, but impose a rather narrower range for the colours of the other items, and in all instances where bright colours are needed for the detail the background should remain quite pale; increasing amounts of background colour would need darker, rather muted colours such as crimson, prussian blue or dark green for the symbols to maintain colour harmony.

Fortunately today the amateur who takes an interest in colour will find plenty of guidance all around him. Book, magazines, posters, films, interior decoration and so on, all make lavish use of colour and the cartographer cannot do better than observe these critically and to take his cue from those examples he encounters which particularly strike his eye.

Colour is perhaps the most widely applicable method of adding eye-appeal to maps and diagrams, but other devices will help too. Symbols of a semi-pictorial, or even a numeral type, as in Figure 34B (p. 109) tend to have rather more interest than simple geometrical ones and Figure 41 illustrates how much more attractive similar information appears when displayed in a more pictorial manner. Both posters were originally printed in bright 'day-glo' colours on a black background, thus appearing quite striking to the eye, yet the second poster was substituted for the first because of the increased interest which it contained. In direct contrast to the simple, yet bold and colourful appeal of this second poster the author once encountered a small leaflet-poster issued to encourage funds for a Diocesan Appeal. The basic device was an annular shape, similar to that of a child's 'oval' railway track, in the centre of which was a representation of a bishop's mitre; the 'oval' band had then been divided, pie-graph fashion, into segments, labelled

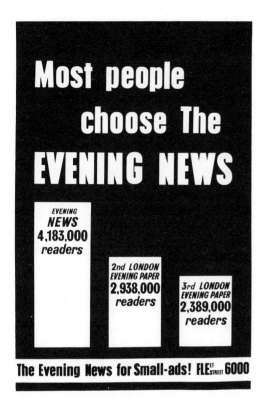

Figure 41   Two posters issued to publicise newspaper circulation. The second poster was substituted for the first because of its greater interest and eye-appeal.

to indicate the way in which the money would be spent, and the whole was printed in black on pale blue paper. In view of the very genuine need to arouse interest in this cause it was unfortunate that on its *visual* merits this small poster seemed unlikely to make any very marked impression. The general design was weak—the oval band was much too thin and unimpressive to act as the main part of the design and this weak effect was emphasised, if anything, by a relatively small bishop's mitre floating aimlessly in the central space; if a strong, bold circular device had been used, the segments could have been larger and more noticeable and the central space could have been small enough just to have contained the mitre which would have acted as a genuine focus for the whole design. The use of a brightly coloured background paper, say a bold yellow, turquoise or medium green would have strengthened the visual appeal of the poster still further and would have rendered it distinctive amongst the mass of church porch notices which seemed likely to form its more usual surroundings.

A feeling of animation and movement is yet another device which will add interest to a map and give more appeal than the presentation of a static picture. In this way, a map involving routes along which traffic is flowing can be improved by incorporating into the design some representation of this movement, such as arrow symbols, rather than mere bands to represent the routes, as in Figure 42. In a rather similar way the use of three-dimensional symbols for points of *particular* interest on an otherwise two-dimensional map will add emphasis and improve what might otherwise be a very 'flat' picture.

Inability to incorporate colour into the design of a map is always frustrating, if only for no better reason than the loss of the potential eye-appeal of bright colours. Unfortunately, this limitation usually imposes other more important and fundamental difficulties into the design, not the least of which are those connected with difficulties of recognition and differentiation. An important asset of bold colours is that they remain distinctive, even where small areas have to be shown, in a way that different types of shading do not; and in techniques such as pie graphs which introduce several variables, often occupying quite small areas, inability to use colour can be a serious handicap. If considerations such as cost or technical difficulties should impose black-and-white-only rendering any shadings used should be designed with particular care so that they remain distinctive when applied to quite small areas. This is not always easy to achieve, but Figure 39 incorporates a range which would be reasonably satisfactory in this respect and which would allow the use of six variables on one map without much danger of confusion.

At no time is the lack of colour more annoying than when a shading range has to be devised to accommodate both positive and negative values of the same thing. The logical choice here would be to progress from white or very pale colours around zero into stronger shades of, say, red for increasingly positive values and blue for increasingly negative ones, and the arrangement is a very satisfactory one. Lack of colour around zero emphasises the 'nothing much happening' range of values whilst the contrast between the strong red and the recessive blue differentiates between positive and negative aspects of the higher values. In black and white this would have to be replaced by two distinct types of shading, e.g. a dot stipple and a line-shading for negative and positive values respectively, but it is difficult to ensure that the two have sufficiently different visual qualities for the negative values to come out less strongly than the positive ones. The only alternative to this is to have an increasing intensity of shading right through the scale of values, with blanks showing the lowest values of all. This makes it difficult to pick out the values around zero although this can sometimes be done by making a major change in shading type, e.g. from dots to lines, at or near that point. It is not always easy to decide which of these two courses is best suited to a given range of figures, and a great deal will depend on which particular parts of the range, if any, need accentuating; examples of both types will be found in Figures 47A and B (p. 147).

Once considerations affecting the design of the map have been completed, there remains the practical task of actually drawing it, and it is at this point, when the final transference of the information to paper must be achieved, that many amateur cartographers feel most at a disadvantage. Consciousness of lack of ability or experience in draughtsmanship all too often leads unnecessarily to a sense of pessimism and frustration—an 'Oh! I'm no good at drawing' or 'I'm not much of an artist' feeling which breeds defeatism, slap-dash work and lack of interest. Naturally the amateur cannot expect to produce work which will compare with that of a skilled craftsman who has been doing the job for years, but on the other hand simple draughtsmanship techniques are not difficult to learn. If the amateur cartographer will not set his sights too high and is

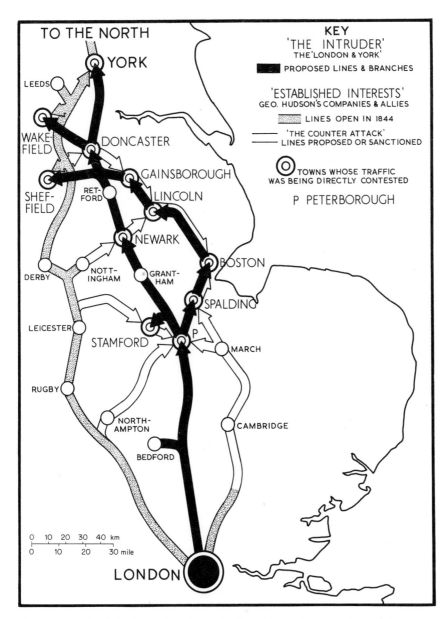

Figure 42 'The fight for eastern England.' The style of this map attempts to convey something of the animation and strife which underlay one of the greatest British parliamentary railway 'battles'. The map is designed to enhance the 'feeling' of the implications of success by the London and York; a *series* of arrows increases animation and these penetrate to the heart of the 'target' symbols whilst others are kept on the perimeter.

prepared to aim at relatively simple, bold results incorporating only a minimum of the more sophisticated techniques (e.g. varied lettering or fine line work) and above all is prepared to take a little time and trouble there is no reason why quite passable results should not be obtained. Contrary to popular belief most draughtsmen are made, not born; their skill is largely acquired by patience and prolonged experience and many people could become much more adequate draughtsmen if they were prepared to accept each new piece of work as yet another opportunity to acquire this skill, rather than as another job to be completed as quickly as possible.

In drawing, as in many other activities, it is true to say that the best results are usually achieved with the best tools for the job and good equipment will always repay the expenditure which it involves. Nevertheless, the person who finds himself required only occasionally to undertake cartographic work will not wish to purchase a very wide range of expensive equipment, and fortunately there is no need for him to do so. Few of the items needed for ordinary draughtsmanship are costly and many of the more expensive ones can often be borrowed or improvised. A good example of an item of equipment of this type is a drawing board which, if available, will form a stable, level and portable base for the work. Alternatively, any large, flat piece of board or plywood, such as a large baking board, will serve the same purpose, provided that it is thick enough, and has one edge straight enough to allow a T-square to be used against it.

A T-square and its complements—a pair of 30°/60° and 45° set squares *are* essential items of equipment for any would-be draughtsman. The T-square should be used against the left-hand edge of the board so that when moved up and down along this edge it can be used to draw horizontal lines in any position. The set squares are used principally to draw vertical lines (and also lines at 30°, 45° and 60° to the horizontal) by moving them along the top edge of the T-square as shown in Figure 43. It is worth while noting that vertical lines are not drawn by using the T-square along the top edge of the board, for

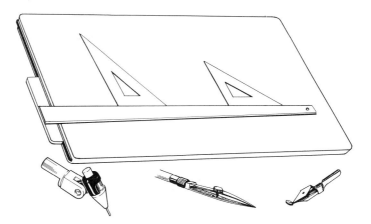

Figure 43   The use of T-square and set-squares with a drawing-board and (*below*) three types of pen commonly employed in cartography: (*left to right*) a 'UNO' lettering pen, a ruling pen, and a dotting pen.

few boards, even of the true drawing-board type, are square enough to give two sets of lines exactly at right angles; it will be obvious that for drawings of any size demanding many vertical lines a pair of large set-squares will be much more convenient than very small ones. The amateur draughtsman will rarely find much use for the rather expensive adjustable type of set-square which can be set to draw lines at any angle to the horizontal.

Of the remaining items needed for the construction of maps none is expensive or difficult to obtain. A good millimetre rule, or 12-in ruler with millimetres[1] marked along the full length, a relatively hard pencil, say 2H, for construction lines, a good quality soft rubber, a small protractor and a pair of compasses will suffice for most work. The hard pencil should be used *lightly* to produce fine construction lines easily erased later, but if used in a rather 'heavy-handed' fashion it may produce small grooves in the paper which are almost impossible to remove completely. Compasses probably offer a wider choice than any of the items of equipment described so far, but the author has found that the simple 'schoolboy' type in cast brass or alloy is usually perfectly satisfactory. In this type the two limbs are joined at the top by a small nut and bolt which can be kept permanently tightened to prevent slipping whilst the pencil 'socket', which is tightened by a threaded brass ring, is generous enough in size to accommodate not only pencils but most types of penholders too, so that the compasses can be used for ink as well as pencil. The main disadvantage of this type of compass for ink work is that it is not possible always to have the writing arm meeting the paper at right angles; more expensive types of compass have the useful refinement of having the lower portion of the writing arm hinged to allow this right-angled contact with the paper to be maintained.

For drawing irregular curves the relevant portions of one or more French curves may be used. These are not expensive and can be bought either singly or in sets. As with set-squares it is sensible to obtain them with a bevelled edge for use with ink as well as pencil. The more elaborate devices for drawing irregular curves will rarely give the amateur draughtsman sufficient use to justify their cost.

The choice of a suitable medium on which to draw the finished map is apt to concern the amateur draughtsman rather less than his equipment or lack of it, yet the choice can be an important one and since most types of paper are cheap enough this is one item in the work where it is easy to obtain the best quality possible. Heavy quality cartridge paper is the medium which most people find both readily available and practicable. The great advantage of the heavier quality papers is that they provide a smoother and tougher surface on which to work and generally withstand wetting, e.g. if water colours are used, more successfully than the cheaper, thinner ones. Their tough surface will be particularly appreciated in facilitating the erasure of any mistakes which cannot be removed by a ordinary rubber (see below).

If the drawing has to be prepared on a transparent medium, for example as an overlay, the usual choice is a good quality strong tracing paper. This has the advantage of being readily available and of allowing easy erasure by scraping, but it is unfortunately quite

[1] In cartographic design the sensible draughtsman will not stick solely to one unit of measurement but will variably adopt metric measure or inches according to which unit is most convenient for his measurements or construction.

unsuited to wetting, which causes troublesome crinkling and is very liable to change its size and shape slightly but noticeably with changes in humidity so that overlays of any size frequently have only a poor quality fit after a time. Avoidance of this stretching can be achieved by drawing on one of the several plastic films which are available, such as 'Ethulon' and 'Kodatrace', though these have the disadvantages of being much less widely obtainable, more expensive and equally unsuited to colour. Such films usually have one shiny side and one matt side, the latter being used for drawing on, but even so the surface may need powdering with French chalk, and Indian ink is liable to flake off if heavily applied; drawing on such a surface needs rather more skill than on cartridge or normal tracing paper. Tracing linen, often used in drawing offices because of its durability in handling, is a similar, difficult medium for the amateur and its use is normally unnecessary for map work.

The medium chosen must be fixed to the drawing board or drawing surface. The amateur is inclined to use drawing pins, which are often at hand, but these spoil both the surface of the drawing and the board and may occasionally foul drawing implements too. The professional draughtsman almost always uses draughting tape, a variety of adhesive paper tape so treated that it can be pulled away from both board and drawing without damaging either. Ordinary transparent cellulose tape should *not* be substituted as almost inevitably it will bring away the surface of the paper if attempts are made to remove it.

Most amateur cartographers are prepared to construct maps in pencil, but have far more misgivings about the final processes of inking in or colouring these. Admittedly liquid media need somewhat more skill and are more difficult to correct if mistakes are made, but even so many cartographers could improve their performance by the application of little more than an increased amount of care and patience, and with only a small amount of practice quite good results can be obtained.

Of all the liquid 'finishing' media Indian inks are the most commonly used, their application to the paper usually being made with some form of pen. These range from pens of the normal 'nib type', such as a crow quill, to pens with special nibs for making particular shapes or lines, e.g. dotting pens, and items such as ruling and lettering pens which are rather different types of devices. The crow quill was formerly much used for map work because of its ability to draw a fine line, greatly curved if necessary, but it is not an easy instrument to use, particularly with regard to maintaining consistency of width of line. Like all pens of the 'nib type' it functions more satisfactorily if the ink is applied with a dropper only to the *underside* of the nib and it needs frequent wiping, for Indian ink dries quickly at room temperature, clogging the pen and making thick lines.

For much of the line work on maps and drawings the professional draughtsman will use a 'ruling pen' which has a 'nib' made of two parallel blades (see Figure 43) between which ink is placed by means of a dropper. The distance between the blades can be adjusted so that the pens will rule lines of any thickness from very fine indeed to about 1·5 mm wide. Ruling pens are not expensive nor particularly difficult to use, at least for straight lines, but they do need to be kept absolutely clean if they are to function successfully, and need sharpening from time to time if they are in constant use. Very sharply curved lines, such as boundaries or rivers, are not so easily drawn with the ordinary ruling pen, although a version known as a 'contour pen' in which the nib is mounted on a swivel to facilitate turning corners is sometimes used. Most amateurs will find, however, that one of the pens supplied for stencil lettering is much easier to use not only for this type of line but also for most line work in cartography. In pens of this sort (see Figure 43) the ink is contained in a cylindrical or conical reservoir and is fed to the paper through a tube which varies in diameter according to the size of the pen. Though supplied principally for lettering these pens are extensively used for line work in cartography today, so much so that they have in many cases almost superseded the crow quill and ruling pen for ordinary work. A range of tube sizes from very small to 4 mm or so in diameter makes it possible to draw lines of a wide range of width; the constant flow of ink from the reservoir produces a consistent line and the tubular 'nib' makes rapid movement in any direction quite easy. Advantages such as this, together with their usefulness for lettering, make a limited range of pens of this kind a sound investment.

Lettering, formerly the bane of the amateur cartographer, need cause little difficulty today. Widely available technical aids, of which the two common types are 'rub down' and stencil lettering, allow the amateur simply and quickly to produce quite 'professional' results in a wide variety of styles and sizes. Both types are easy to use. In the 'rub down' variety, such as the many 'Letraset' alphabets, self-adhesive letters are transferred by rubbing from a pre-printed sheet onto the map, and the range available includes a multitude of shadings, stipples, symbols and even colour tints as well. Cut to shape and then applied to the relevant portions of maps these 'rub down' shadings and stipples can save much tedious, repetitive work, at the same time giving excellent results, particularly with devices such as even stipples which are difficult and tedious to produce by hand over large areas. (Alternatively hand drawn shading can be quite effective and satisfactory, provided that care is taken to draw guide lines in first in the manner already described in Chapter 2.)

Well known makes of stencil lettering are 'UNO' and 'Standardgraph' and here too stencils for drawing graduated ranges of symbols, e.g. squares, triangles, circles, crosses are also available. When lettering with these stencils the lower edge is usually allowed to slide along a T-square or straight edge thus maintaining horizontality as the stencil is moved to bring the next letter into its required position.

If stencils or rub down alphabets are not available, lettering may have to be done by hand and in this case most amateurs would benefit by the generous provision of guide lines and the use of a set-square to ensure that all 'uprights' are truly vertical or slope at the same angle if an italic style is used. Variations in style in freehand lettering can be produced by the use of different shaped nibs, of which a wide range is available. Those supplied under the trade name of 'Pelikan Graphos' include nibs to produce a comprehensive range of styles and in Britain cards of nibs of different sizes to produce a particular type of lettering and sold by the firm of Wm. Mitchell are frequently seen in stationers' shops. If attempts at freehand lettering prove unsatisfactory another solution is to type the necessary names on gummed paper, cut them out and stick them firmly to the map. This solution will, of course, place all names on the map in a white 'window' which may or may not be an advantage; the results may furthermore be rather mechanical in appearance, but this is probably to be preferred to a map rendered untidy by poor hand-lettering.

Perhaps one of the main reasons for the amateur's mistrust of liquid media, as opposed to pencil, arises from the greater difficulty in correcting any mistakes which may be made. Naturally, mistakes are better avoided altogether, but even the professional draughtsman will err occasionally and the important factor is not so much not to make mistakes as to know how to correct them when they are made. Any mistakes which will not respond to an ordinary eraser are usually removed by scraping the ink or colour off the surface of the paper by means of a sharp, straight edge such as a razor blade scraped along the paper itself or at a very small angle to the surface; if the blade is held at a larger angle the end of the cutting edge will tend to dig in, breaking through the working surface of the paper and roughening the looser fibres underneath. With cheap, thin papers which have little working surface this scraping process can only be used with caution, but with harder, thicker paper, or on tracing paper, a surprising amount of erasing is possible by this means, and on the very best quality paper such as hand-made Whatman or on Bristol Board[1] the surface is hardly affected. After erasing, it is usual to smooth out any roughness which may be present by rubbing the erased portion with some hard, smooth object. Most professional draughtsmen carry a small piece of bone for this purpose, but other surfaces, e.g. a rounded portion of a perfectly smooth, plastic penholder, can be substituted.

Really drastic mistakes such as a very large blot or smear or an incorrectly drawn section too large to erase satisfactorily may need more serious treatment if the whole map is not to be redrawn. The drawing should be laid on a smooth surface with a piece of the same paper placed underneath the area which contains the mistake; this area should then be cut out of the map with a very sharp knife, ensuring that the blade penetrates cleanly *both* thicknesses of paper. If the spoilt portion is now removed a blank piece of the same material, and of exactly the right shape, is available to fill the hole and can be held in place by adhesive tape on the back of the map. For drawings which are to be reproduced as line blocks to accompany text, erasure is relatively easy, for the mistakes can simply be painted out with an opaque white substance known as 'process white'. Process white does not show up when the drawing is photographed in the making of the line-block and the obliterated error is entirely invisible on the finished representation.

Whilst Indian inks are perfectly suitable for coloured line work their waterproof quality makes them difficult to apply evenly over large areas and, for this purpose, some other colouring medium must be used. Large areas of pale colour can be satisfactorily achieved by using water colours or stick inks made up with water and generously applied with a large brush, but unfortunately it is not easy for the amateur to imitate the large areas of solid *bright* colours which are responsible for much of the attractiveness and eye-appeal of printed maps and diagrams. The lithographic process which is normally used in printing these maps is particularly adapted to this kind of effect, but has no easy parallel in 'hand' methods of applying colour. Poster colour will be found to be the most suitable medium for producing large areas of opaque bright colour and though the results are usually perfectly satisfactory, most people find it a slow medium in which to work and one which is apt to flake off if applied too heavily and roughly handled afterwards. For slightly less dense patches of colour a wide range of non-waterproof inks is available.

---

[1] A type of thin, white card with a very smooth surface, often used where a very durable drawing is required.

Their non-waterproof nature makes them rather easier to apply evenly over large areas than Indian ink, but it is notoriously more difficult to achieve satisfactory results with some colours than others and a trial run on a piece of scrap paper is often advisable before applying the colour to the main map. Because these non-waterproof inks are not opaque all construction lines and unwanted detail must be removed before they are applied.

It is unfortunate that many people regard crayon as rather a 'childish' or second-rate colouring medium. Although not much used by the professional draughtsman for his finished work it has several advantages for the amateur. Its 'pencil' nature gives confidence, it is quick to use and is capable of producing a wide range of colour intensity, matching that of both poster and water colours combined, as well as being adapted to producing 'shading-off' or gradation in colour which is not easily achieved by the amatuer in any other medium. The greatest difficulty with crayon arises when large areas have to be shaded evenly, but the secret here is to employ a good quality crayon with a *soft* lead and work gradually over the area to be covered with a gentle circular motion, treating relatively small areas at a time. Intensity of colour over large areas is more uniformly achieved by producing a lighter colour first and going over the area again rather than trying to build up the full intensity from the beginning. Shading with a side-to-side motion, which is the method normally adopted by amateurs, will tend to produce variations in intensity, making it difficult to obtain the evenness of tone which is desired. If several shades of one colour are needed, e.g. in a graduated range of shading, it is perhaps best to use solid colour only for the highest value and obtain the other shades by building up colour intensity with line shading; it is not easy to maintain several recognisably distinct shades of *solid* colour, using the same crayon for each. It is probably true to say that if the finished map is relatively small the amateur can achieve speedier and more striking results working with crayon than with most other coloured media, but unfortunately the shiny surface produced by most crayons means that ink cannot be applied successfully to a crayoned surface and any lettering must therefore be either hand-drawn in crayon or pencil, or typed and stuck on as described above.

The help and advice given in the latter part of this chapter are, quite obviously, intended for those with little experience of draughtsmanship. The simple, relatively easily acquired techniques and skills which are described here should encourage even those with little experience to attempt to produce neat, effective cartographic work, although naturally the results achieved at first will fall far short of those produced by the more experienced draughtsman who will already have encountered the different media and established his preferences amongst them. Even so it should be remembered that there is far more to a map than a neat appearance and good draughtsmanship. Good draughtsmanship, however desirable, must take its place along with attention to sources, sensible use of material, choice of method, and general design as merely one factor among many which must be incorporated into the construction of any map or diagram. On its own, however glib or professional it may appear, it will not automatically produce a good map and the person who has carefully weighed these other features need not be hypersensitive that limitations in his draughting ability will seriously reduce the value of his work.

7 Some worked examples

In the foregoing chapters the essentially compound process of devising a statistical map or diagram has had to be analysed into its component parts so that a set of principles, applicable to each of these could be evolved. In practice, of course, the difficulties presented by each aspect of the situation cannot be resolved separately, but must be approached in the light of their interaction one with another if a pleasing, balanced and efficient map is to result. Using worked examples of statistical situations this chapter seeks to restore this balance and, by presenting each problem as an entity, gives to the reader a limited idea of the considerations involved in viewing such a situation 'in the round' rather than from a narrow specialised point of view.

The examples used have been chosen principally because they are illustrations of certain commonly recurring types of statistical situation (they are in fact described under 'type-headings' rather than as individual problems), but also because it is precisely these kinds of distribution which the amateur cartographer so often finds himself having to illustrate.

Example 1   A distribution concerned essentially with a single variable. The distribution of the chemical industry in England and Wales, 1951 (Figure 44)

Of all types of distribution which have to be mapped, that dealing with a single variable is perhaps the most simple and also one of the most frequently encountered. Let us take as an example an attempt to show the main features of the distribution of the chemical industry in England and Wales in 1951.

Source of information
This must be considered first, as the information it yields will inevitably form a limiting framework to the design of the whole map. Fortunately, an excellent and complete source is available in the Industry Tables of the 1951 Census of England and Wales, the total numbers employed in 'Chemicals and Allied Trades' being given for each Local Authority area (i.e. County or Municipal Borough, Urban or Rural District). As defined in the Industrial Classification used in the Census the term 'Chemicals and Allied Trades' includes a range of activities rather wider than those which the layman often thinks of under the heading 'chemicals'. For towns of more than 50,000 population the total employed in the whole industry and in each of seven main subdivisions is given; these are

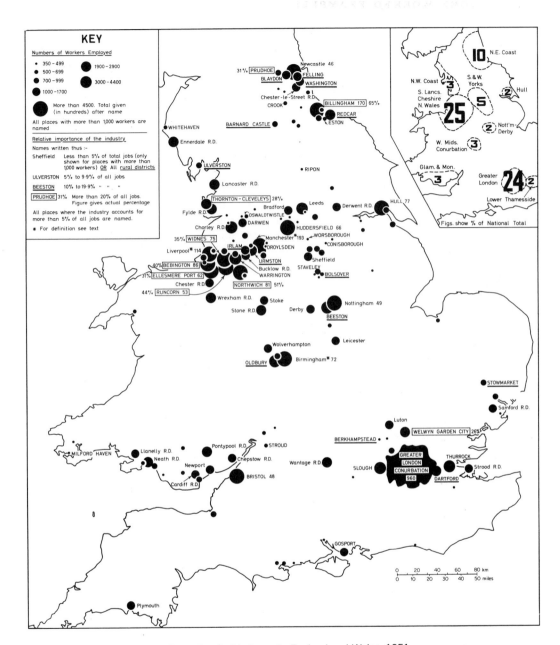

Figure 44 The distribution of the chemical industry in England and Wales, 1951.

Source: Registrar-General, *Census of England and Wales, 1951*, Industry Volume; London (H.M.S.O.), 1957.

(*a*) coke-ovens and by-products; (*b*) chemicals, dyes, explosives, fireworks; (*c*) pharmaceutical, toilet preparations, etc.; (*d*) paint and varnish; (*e*) soap, candles, polishes, ink, matches; (*f*) mineral oil refining; (*g*) other oils, greases, glue, etc. Unfortunately for smaller authorities the only subtotal is that for groups (*a*) and (*b*) *combined*.

Aspects to be considered

The principal impression of the distribution of the industry throughout the country must inevitably be derived here from the variation in the distribution of workers employed in it. There is no uniform source of detailed information based on production, nor is it easy to see how one could be devised to accommodate the various end products of so wide a range of activities as those specified above. *Some way of presenting quantitative aspects of employment in the industry at the various places where it occurs is therefore a prime requirement.*

A glance at the figures suggests that other aspects might be relevant in obtaining a general picture of the industry, e.g. there is considerable variation of industry within the main industrial class and even a cursory inspection reveals that there are at least two basic types of *place* involved. These are (*a*) the 'chemical town', where chemicals form one of the main sources of employment and assume, therefore, unusual importance, and (*b*) the town where chemicals are merely one industry among many, even though the total number employed in chemicals is quite large. Contrasting examples of these are provided by Runcorn U.D., where the chemical industry provided 5,304 jobs or 44% of all employment within the Urban District, and Leeds where a slightly smaller total of 4,227 jobs in the chemical industry accounted for a relatively insignificant proportion ($1\frac{1}{2}$%) of the total employment. *Thus some means of indicating the importance of the industry in the places where it occurs would be desirable.*

General statistical considerations

The territorial distribution of the industry and the range of numbers employed are both very wide indeed. Few local authorities are completely without people employed in 'chemicals', though in many cases the totals are very small, often less than 20; on the other hand the 'giants' give figures of several thousands, and totals of several hundreds are not uncommon. It seems fairly clear from this, that in a national survey of this kind which must attempt to emphasise the essentially important aspects of the distribution it would be pointless to clutter the map with many examples of places where the industry is of little importance, and after inspecting the figures in several typical localities it was decided to omit all authorities where the number employed fell below 350. This figure was decided upon quite arbitrarily, partly by the general run of the figures and partly from individual knowledge, the main aim being to ensure that no place which was known to have an important chemical works appeared to be excluded. A more precise solution to this problem could have been obtained by means of a dispersal graph, but this would have been unnecessarily tedious to prepare in view of the very large number of occurrences involved. Even so, this whittling down left some 200 or so localities to be represented and the range involved was still very wide, running from 350 jobs to more than 10,000.

To avoid the kind of statistical anomalies mentioned in Chapter 3 it was decided to amalgamate at least the figures for Manchester, Stretford and Salford; Liverpool and Bootle; Newcastle and Gateshead; and Birmingham and Smethwick. To avoid congestion and to allow the London area to stand out in its true perspective, figures for places within the Greater London Conurbation[1] were included as one unit. The great majority of the figures used related to urban authorities, which, though technically 'areas', were small enough to be reduced to and regarded as 'points' on the finished map; and for these areas it was also meaningful to express the total of chemical workers as a percentage of all employment. The remaining figures appeared to present rather more of a problem. They related to Rural Districts which were often of such a size that it would have been impossible to regard them as 'points' had it not been for the fortunate fact that in the great majority of cases it seemed likely that the figure was derived largely from only *one* major establishment which could usually be identified (e.g. the Harwell Atomic Energy Establishment obviously accounted for the great majority of the total of 2,796 'chemical jobs' in Wantage R.D.) and a point symbol located at that particular place would serve perfectly well. *The basic distribution, therefore, though technically relating to areas could in fact be treated as a series of 'point' values.* Unfortunately, with respect to *composition* the rural areas could not be dismissed so lightly. A single large works at one end of a Rural District might be so inaccessible as not to affect potential employment at the other extremities of the area, and to express 'chemical jobs' as a percentage of something which in this respect scarcely functioned as a unit seemed pointless. In many cases too, the works, though technically in Rural Districts, were, in effect, close to urban authorities and probably drew more of their workers from these towns than from the remaining parts of the same Rural District. Thus the 2,092 'chemical jobs' in Lancaster R.D. were largely concentrated at the oil refinery at Heysham, only a short distance from the towns of Morecambe and Lancaster; to express this figure as 37% of all jobs in Lancaster R.D. seems simply to refer to a fictional situation, entirely unrepresentative of conditions in Lancaster R.D. generally. *For sites in rural areas, therefore, no attempt is made to express the relative importance of employment in the chemical industries* although it can be assumed that in a rural area *any* large factory forms an extremely important local source of employment.

Choice of method and details

The basic choice of method was never in doubt. With almost 200 values ranging in size from 350 to 10,000 a series of quantitative symbols was needed, either strictly proportional individually or grouped into size categories. The latter solution was decided upon principally because it was felt that with so large a number of symbols the map reader needed a certain amount of guidance with information sorted out into different degrees of importance. A dispersal graph of the values established the most suitable division points. It will be noticed that the larger categories cover a relatively wide range of values (the upper limit is frequently 50% or more in excess of the lower one), but this is almost inevitable as values on the dispersal graph 'thin out' and indeed above 4,500 occurrences become so dispersed as almost to preclude the possibility of grouping. In this last group of 'giants', which contains places all of which could be said to be amongst the main centres of the industry in the country, there is indeed so wide a range that some other means of indicating quantity is needed as well. In defining a suitable size for the symbols an attempt was made to keep some aspect of the actual quantities in view, and the diameters used are

[1] Defined in the *Census of England and Wales, 1951*, Industry Volume.

in fact close to those which would have represented a value half-way between the extremes of each group if a strictly proportional symbol had been used. The series does, however, contain two marked 'steps' in symbol size—namely, between the third and fourth sizes used and between the sixth and the seventh. These are introduced partly to break up the range and make recognition easier, but principally because each marks a major change in the extent of the groups, the first three symbols covering relatively small ranges and the last one (the seventh) an extremely extended range. To cover the extended range of the seventh group all symbols also carry the actual numbers employed (in hundreds) after the name. Whilst it would have been more satisfactory to have placed this total either adjacent to, or within the actual symbol, congestion in some areas, notably South Lancashire, rendered this impossible.

Having overcome the problem involved in establishing the primary distribution the next stage was to consider how best to add to the map a second aspect, indicating the relative importance of the industry at each place. One possibility here was to shade the symbols differentially, but this was not used since a uniform solid symbol is important to give the maximum visual emphasis to the principal distribution. Instead, the idea of introducing differentially written names was adopted. Names are easily written in a number of different ways to indicate various categories, as well as serving as a useful form of identification. Not all locations shown on the map are named; in an 'all black' map these would soon fill the map and destroy the impression of the first distribution so that names are only given for those areas where more than 1,000 people are employed in the chemical industry or where the industry accounts for more than 5% of all jobs. Names in the latter group are presented in three ways, according to whether the percentage exceeds 5, 10 or 20, and in accordance with the principle established above this aspect of the figures was not presented for Rural Districts. The limits of 5, 10 and 20% were quite arbitrary and were chosen because in the majority of cases they could be determined by inspection without need for further calculations.

The possibility of introducing a third aspect on to the map to illustrate variation in the type of industry was considered, but rejected. Had colour been possible it would not have been difficult to achieve this by using the same symbols or names differentially coloured. In black and white the counterpart of this would have been differentially shaded symbols and this had already been rejected. The alternative was to incorporate information of this type in a separate inset map where different symbols (non-proportional) could have indicated towns where various branches of the industry were dominant. Unfortunately, there was room for only one inset map and it was decided to utilise this space for purposes which essentially completed the kind of information given in the main map—namely, a summary diagram which shows the general national distribution of the industry. With so many individual symbols involved it is difficult for the map user to judge precisely what the total effect of these in any one area amounts to, especially in comparison with other areas; to overcome this difficulty the inset map expresses the total employed in certain of the main concentrations as a percentage of the total employed nationally in the industry.

General conclusions derived from Figure 44
(*a*) The industry is widely distributed with certain locational types commonly occurring—

namely, (1) in the main conurbations and large towns and (2) in coastal localities. Three concentrations in particular stand out above all the rest. Greater London and the South Lancashire-Cheshire area each have about a quarter of the total employed nationally in the industry, and the Tyneside-Tees-side area accounts for a further 10%. Other areas are not individually important on a national scale though sizeable numbers of workers may be involved.

(b) Areas with over 1,000 persons employed in the industry are frequently encountered, but with the exception of the Mersey area the largest totals are usually in the big cities where the industry is *relatively* of small importance. Generally speaking, it is often in the smaller town, where numbers involved may be less than 1,000, that the chemical industry forms a substantial proportion of local employment. South Lancashire, where large numbers are employed in medium-sized towns, is exceptional and in that area chemicals account for a considerable proportion of local employment.

Example 2   A distribution concerned essentially with a single variable, but with major statistical complications. Population change in the North Riding of Yorkshire, 1931–1951 (Figure 45)

One of the most serious defects of published statistics is that they are, at times, capable of being quite unrepresentative of a *detailed* situation. This defect was examined in broad outline in Chapter 3 where examples illustrated how the use of different bases for representing population change from 1931 to 1951 in the North Riding of Yorkshire could produce widely varying results. The end-products of this experiment were presented there in Figure 21 (p. 74), but it was pointed out that even Figure 21C, the best of the results, was more expedient than accurate and the further consideration of the figures which was promised at that time will now be attempted.

Source of information
The same source will be used here as in the compilation of Figure 21—namely, the Yorkshire East and North Ridings volume of the 1951 Census of England and Wales. This provides figures for the 1931 and 1951 populations for each administrative area, including figures for individual parishes in rural areas. Figures are given for both total population and 'population in private households' (i.e. excluding people in institutions and communal establishments such as hospitals, hotels, service camps, approved schools, etc.), although the figures for this latter category apply to 1951 only and the corresponding 1931 figures had to be obtained from the relevant volume of the Census of that date.

Aspects to be considered
Unlike the previous example only one major aspect of the figures will be treated. The aim is to produce a map which will establish a picture of differential aspects of population change throughout the area and for this purpose *the change must be expressed relatively, e.g. as a percentage increase or decrease, rather than numerically.* Because of their very different nature and fortunes during the period under study *it would be desirable for the map to differentiate clearly between urban and rural areas.*

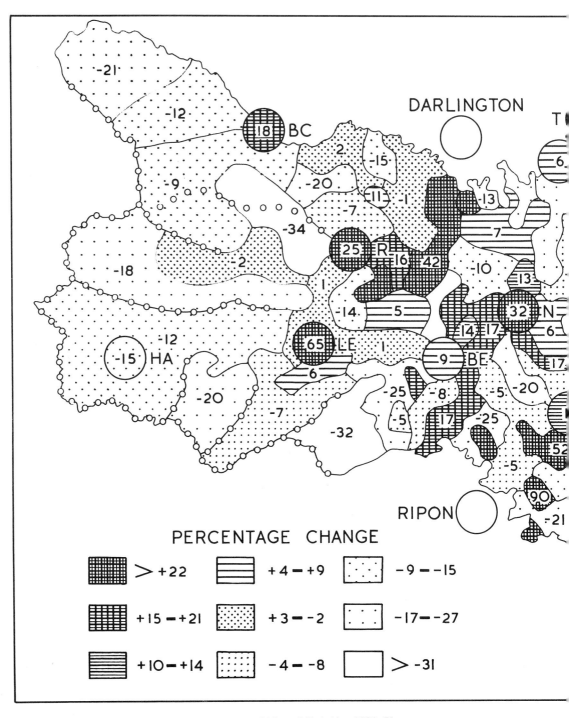

Figure 45 Population change in the North Riding of Yorkshire, 1931–51.

Source: Registrar-General, *Census of England and Wales*, 1931 and 1951. County Volume. Yorkshire (East and North Ridings); London (H.M.S.O.), 1954.

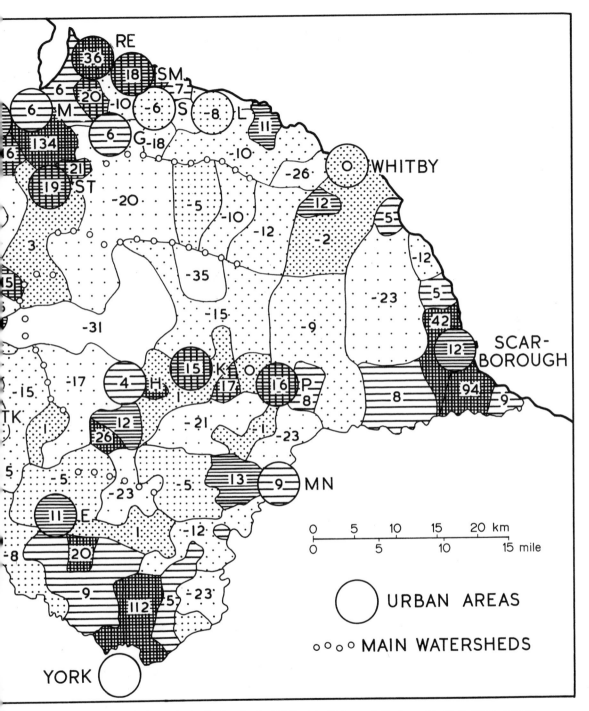

Key to towns shown. BC Barnard Castle-Startforth: BE Bedale-Aiskew: E Easingwold: G Guisborough: H Helmsley: HA Hawes: K Kirkby Moorside: L Loftus: LE Leyburn: M Middlesbrough: MN Malton-Norton: N Northallerton-Romanby: P Pickering: R Richmond: RE Redcar: S Skelton: SM Saltburn-Marske: ST Stokesley: T Thornaby: TK Thirsk-Sowerby: Y Yarm.

General statistical considerations

The hard core of the problem here has already been touched upon in Chapter 3. The published figures, which relate to the component Rural Districts as a whole have already been shown frequently to be meaningless, the administrative boundary lines often arbitrarily containing areas of very different types so far as their recent population history is concerned. Unfortunately, breaking down the figures into individual parishes, which seems at first sight the most likely way of overcoming this difficulty, would not be a particularly satisfactory solution for it was also pointed out in Chapter 3 that the majority of the parishes within this county are so small in total population that random variations could easily produce figures just as arbitrary as those for whole Rural Districts. Even if this had not been the case the result of dealing with individual parishes would have been a map so patchy and fussy that it would have been impossible to deduce from it the dominant trends suggested by the mass of detailed information. The only solution, therefore, is to seek some grouping of parishes more logical than the administrative one, large enough in total population and uniform enough in aspects of change to allow a reasonably representative figure to be given for it.

To do this the individual figures for the change in all parishes were first calculated and plotted on tracing paper placed over a map indicating parish boundaries. Examination of the 'run' of the figures on this map facilitated the grouping of parishes showing reasonably similar trends, and a figure for the change for the whole group was obtained by calculating the total population of the component parishes, in 1931 and 1951. Urban areas, including any necessary adjoining parishes, were always treated separately, as was any place with a population of over 800. This last figure was established from a dispersal graph of parish population size and was chosen because it represented a type of big village, quite unlike the normal village in the area, and moreover seemed large enough at least to allow cancelling out of most of the random errors.

For the rural areas the method worked quite well. To avoid leaving the odd small parish on its own, exceptions to the general group-trend had sometimes to be included, but in only a very few cases did the figures for individual parishes differ by more than 4% from the group figure; and the total population of the group rarely fell below 800 (it was often two or three times this figure) so that once again the effect of random variations would be considerably removed. Another important feature in obtaining representative and comparable values throughout the areas was the decision to use the figure relating to *population in private households* and *not* total population as the basis for the calculation of change, at least for all rural groups and towns with below about 10,000 population. In towns larger than this the changes calculated from either set of figures never differed by any significant amount and in many cases with these larger towns boundary changes made it impossible to obtain comparable figures for the population in private households in 1931 and 1951. For smaller towns and especially for some rural parishes the two figures differed very much indeed, perhaps the most notable case of all being the small town of Leyburn where the two results were as follows:

	1931	1951	Change	Change %
total population	1,440	1,281	−159	−11
population in private households	728	1,202	474	65

The remarkable difference being solely due to the presence of a temporary military camp in Leyburn on census night in 1931. It is the presence of particularly large numbers of servicemen on aerodromes and Army camps which is responsible for *some* of the worst anomalies in the figures for whole Rural Districts given in Chapter 3, but examination of detailed figures for individual parishes shows that less obvious 'institutional' factors are responsible for misleading results being obtained from the total population figures in a considerable number of cases. To obtain a figure which purports to represent change in a parish but which in fact reflects not conditions generally but merely the fact that four more old people now reside in a home for the aged, and then to use that figure to make comparison with conditions in neighbouring areas is surely a pointless task. Comparison needs a figure based on relatively uniform conditions and one derived from the 'private households' total will provide this.

Choice of method and details
The modified statistics prepared as described above related to a series of areas of varying size and, as in the previous example, there was no difficulty in choosing a satisfactory method; some form of shading technique was a natural choice. The main secondary consideration here was a statistical one. The representative figures for each area had been obtained only as a result of considerable calculation and it seemed pointless to allow this work to be lost again in the broad shading-groupings of the normal shading map, so that some means of retaining accurate representation of the value was desirable. As a first choice proportional shading, as in Figure 16F (p. 52) was envisaged, but lack of colour proved a serious disadvantage. With two colours, one for increase and one for decrease, and by slightly modifying the method so that the colours became solid at only moderately large values (everything above that value being shown in solid colour representing an unusually large increase or decrease) a reasonable vertical scale for the shaded bands would have been obtained and the method might have worked well. Unfortunately this technique was impracticable in black and white. The idea of replacing the two colours by two tones, e.g. solid for increase and dotted for decrease, seemed unlikely to differentiate satisfactorily between positive and negative changes and to use one tone only, so that the range of values had to include both negative and positive ones, would have given too restricted a vertical scale. Instead, recourse was had to a normal type of shading map with shading-groups established from a dispersal graph, but with the actual value superimposed wherever space permitted this.

Differentiation between urban and rural areas was achieved by using a uniform circular symbol for all urban areas (whether administratively Urban Districts or not) though in all other cases area limits were taken from actual parish boundaries. Urban areas are thus quite easily recognised on the map and the scatter of circular symbols (which are named) provides a useful framework to help the map user locate the various subdivisions of the rural areas. To assist this latter process even further within the rural areas themselves, the main watersheds are marked unobtrusively, enabling the Pennine Dales or Eskdale west of Whitby, for example, to be located easily without the need for adding more names to the map.

General conclusions derived from Figure 45

The map emphasises above all things the detailed complexity of population change in this area and destroys absolutely the idea of facile approximations based on figures for entire Rural Districts. In particular, certain important areas of depopulation stand out, notably the centre of the Vale of Pickering and the remote valleys enclosed by moorland in both the Pennines and the North York Moors, although even in these cases conditions are far from uniform over large areas. Other factors which emerge are the unusually complicated pattern of change in the northern portion of the Vale of York, and, outside that area a tendency for the noticeable increases in population to be confined to urban areas and to the quite restricted areas of 'suburban expansion' around York, Scarborough and Tees-side. In the case of the three last mentioned areas the *numbers* involved are quite large and are particularly apt grossly to distort the figures for much wider areas if returns are made for entire Rural Districts.

Example 3   Assessing the effect of the interaction of two related components within the distribution of one main variable. Analysis of the population change within the central portion of the West Riding of Yorkshire, 1931–1951 (Figures 46–49)

In the previous example population change was considered quite simply as a single variable. The figure given was that of total change and no attempt was made to investigate any factors which contributed towards that figure. In practice it is well known that this population change is the result of the interaction of two factors, natural increase and migration, and that quite similar total changes in two areas may result from the interaction of these factors in very dissimilar fashion. Natural increase is the population change which would result in any area from the excess of births over deaths (in some areas in Britain excess of deaths over births is occasionally encountered), and this figure in any area and for any intercensal period can be calculated from the returns of births and deaths made to the Registrar-General. The term migration is self-explanatory, but somewhat misleading. What is implied is *net* migration, i.e. excess of immigrants over emigrants or vice versa in any particular area, and this figure is usually obtained as the difference between the total change and natural increase thus:

	Natural increase	Total change	Assumed net migration
area A	6·0%	16·0%	10·0%
area B	17·0%	9·0%	−8·0%

The task which is being attempted is a study of population change within an area not simply in terms of total change but also by analysis into these two component parts, with areas classified differentially on this basis.

Source and statistical considerations

Information on all three aspects, total change, natural increase and migration for the period 1931–51 is available, for each administrative area, in the West Riding volume of the 1951 Census of England and Wales. For urban areas this can be considered relatively satisfactory, but for rural areas, where only composite figures are available for each Rural District, serious misgivings arise. It is disconcerting to find, for example, a figure of $+17\cdot0\%$ given for migration for the whole of Wharfedale R.D. within which individual parishes show frequent *total* changes of -10% to -20% or more with implications of possible *negative* migrational changes in excess of these figures. Unfortunately, figures for natural increase and migration are not available for individual parishes and since it is obvious that, as in the last example, figures for entire Rural Districts lump together individual parishes of greatly dissimilar type some attempt at a breakdown of the published figures is essential.

It is not easy accurately to provide this breakdown, but an approximation which is at least superior to the unadjusted values can be obtained starting from the total change which occurred in *groups of parishes* after the manner described in the previous example. Once this total change is established, if the same figure for natural increase is used as for the whole Rural District, a new figure for the net migration from each group of parishes can be obtained as indicated above. The validity of the results is dependent on the reliability of the assumption that the figure for natural increase remains relatively constant throughout a whole Rural District. Fortunately, the figures for natural increase, when plotted separately as in Figure 46A, do show a fairly gradual and even trend throughout much of the map, so that rapid variations within a small area are unlikely. Figure 46A also suggests the possibility of calculating intermediate values, but the fact that within each group of parishes two or three intermediate values might have to be used would lead to excessively complicated calculations; and as this would still not entirely remove the element of approximation, it was decided generally to adopt a uniform natural increase figure for the whole Rural District. An exception to this rule was made in the case of Doncaster, Hemsworth and Osgoldcross Rural Districts which occupy much of the area between Doncaster, Barnsley and Pontefract. The trend of values in Figure 46A suggests that the remarkable gradual increase towards the south-west corner of the map is in some way related to the increasing importance of coal mining within the area and in each of these three Rural Districts mining areas are intermixed with enclaves more noticeably in agricultural use. Accordingly, for the agricultural areas in these districts a figure for natural increase roughly comparable with that in adjoining agricultural Rural Districts was assumed, and from this the resultant natural increase for the rest of the Rural District could be calculated.

As an example of the discrepancy between published figures for the whole Rural District and the figures actually used, the detailed breakdown is given below for that portion of Wharfedale R.D. contained within the map. Though the figures are necessarily only *estimates* it cannot be doubted that they are to be preferred to the highly unrepresenta-figure for the whole Rural District.

	Total change	Natural increase	Assumed migration
Wharfedale R.D. (whole)	25·2%	8·2%	17·0%
1 Parishes north-west of Otley	−16·8%	8·2%	−25·0%
2 Newall-with-Clifton, adjacent to Otley	9·6%	8·2%	1·4%
3 Parishes north-east of Otley	−2·3%	8·2%	−10·5%
4 Parishes between Otley and Leeds	50·4%	8·2%	42·2%

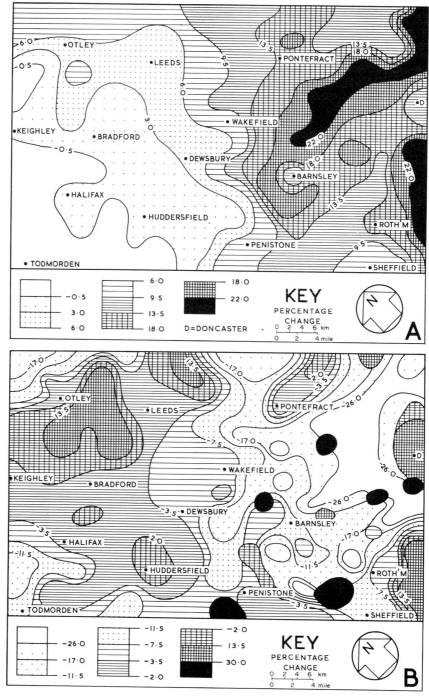

Figure 46 Components of population change in part of the West Riding of Yorkshire 1931–51. **A** Changes due to natural increase. **B** Changes due to migration.

Source: Registrar-General, *Census of England and Wales, 1951*, County Volume, Yorkshire (West Riding); London (H.M.S.O.), 1954.

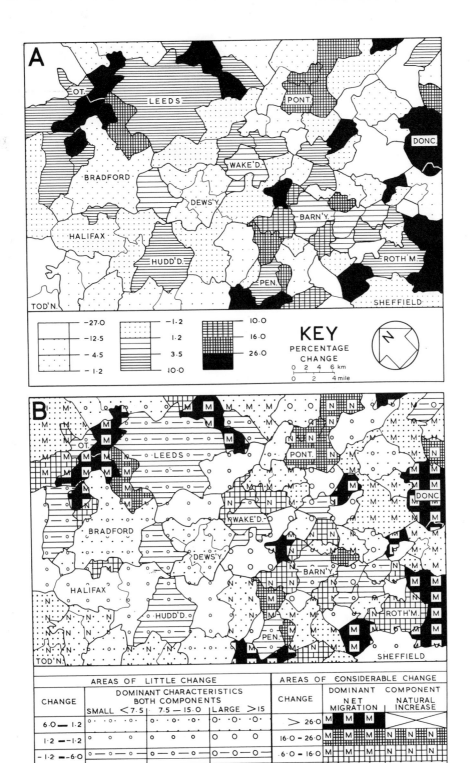

Figure 47 Population change in part of the West Riding of Yorkshire, 1931–51: **A** By simple shading technique without consideration of components. **B** By shading technique adapted to show dominant components as well.

Source: As Figure 46.

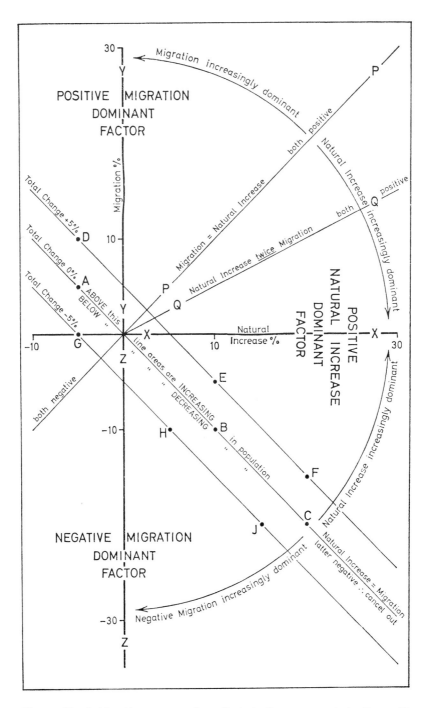

Figure 48 A 'sketch' scatter graph to illustrate the components in Figure 46, showing significance of situation on certain lines or in certain areas of the graph.

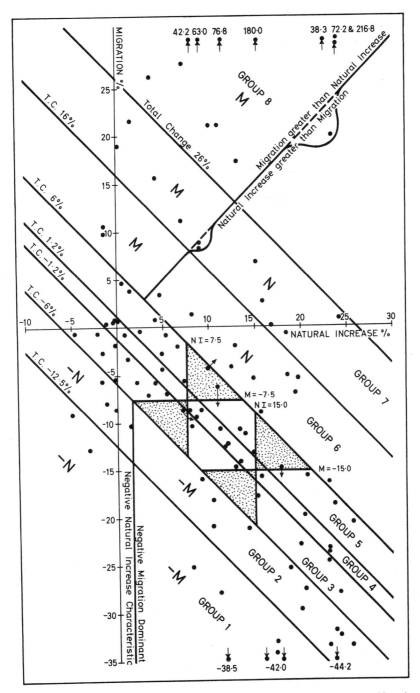

Figure 49   Actual scatter graph of the values used in constructing Figures 46 and 47.

Source: As Figure 46.

No modifications were made to any other figures except in the case of Doncaster County Borough whose anomalous boundary was 'redrawn' to include the Urban District of Bentley-with-Arksey and the parish of Sprotborough, Doncaster's obvious suburbs lying north of the River Don.

Choice of method and details

In assessing the interaction of two related components within one main variable the problem presented here is the most involved which has been considered so far, and as was explained in Chapter 4 there are at least two possible lines of approach. The first of these is what might be termed the 'obvious' or simple one and consists in presenting the final picture through *three* maps, one to show the overall change and establish the basic pattern under investigation, and two more, one each to illustrate natural increase and migrational change. The last two maps would provide further illustrations of trends and background information with which to reassess the implications of the first map.

This approach, with its relatively simple cartographic problems, has already been criticised in Chapter 4 since the ultimate aim of the study is really achieved not carto-graphically but *mentally*, being left to the perception of the map user. The second line of approach which presents much more serious cartographic problems, requires some tech-nique which shall contain *on one map* a representation of both change *and* its component elements. Let us examine each of these approaches in turn.

The three maps needed for the first approach do not present any serious difficulties other than the need for better statistical information regarding rural areas already men-tioned. In each case the figures used relate to areas and for the map of *total* change a well-designed shading map (Figure 47A) seems most likely to fulfil the requirements of telling as straightforward and specific a story as possible, indicating the various areas used, their location and the amount of total change. Originally it was hoped to include additionally within each area, in figures, the actual amount of change, but within the limitations of black and white technique this proved impossible; each value would have had to be placed in a 'window' which would in turn have given rise to difficulties in naming sufficient areas to provide the index of location which was just as necessary as a representation of values. With colour techniques the possibility of over-printing names and figures on differently coloured shading would have made it possible to incorporate these desirable additions quite easily. As a substitute for this absolute precision a fairly large number of shading tones (9) was used and each of these was designed to cover as narrow a range of values as possible, the actual limits being determined in the usual way from a dispersal graph.

An alternative, precise solution would have been some form of proportional shading, but on the whole the range of values was too wide and more particularly many of the areas were too small to make this method very feasible. As in the previous example, the lack of colour differentiation between positive and negative values had to be overcome by using increasing intensity of shading, the very lowest values being left blank and the position of zero being emphasised by a change from dot to line shading close to that value (actually at $+1 \cdot 2\%$).

The two supplementary maps serve rather different needs and demand slightly different treatment. Unlike the first map, which had to be as definite as possible, these would be more suitably designed to show *trends* within the area so that their message could be fairly quickly assimilated by the map user and mentally stored for use in assessing the first map. Of course it would be perfectly feasible to present the information here by exactly the same technique as was used for Figure 47A, but it was felt that this would only over-whelm the map user with more detail than he could comfortably handle and that a combination of one fairly precise and two fairly generalised maps was much more manage-able. For the illustration of trends among a series of plotted values relating to areas, the isoline map is an obvious choice and this technique has been used for both maps. The first map (Figure 46A), relating to natural increase, presented no problems and the isolines have been drawn by interpolation as described in Chapter 2, plus a certain amount of guidance from the run of the local authority boundaries. The figures for migration were not so easily dismissed, however. Unlike natural increase, trends were not always gradual and in particular awkward 'islands' of quite considerable increase were often found amidst much lower or even decreasing values. To have introduced these in the normal way would have involved tedious building up to the required value by numerous closely spaced ring isolines; this seemed both messy and unnecessary and it was decided to treat these instances as anomalies in the general pattern by breaking the normal run of the isolines and simply inserting a high value 'blob' across them, much in the same way that isolated unconformities are marked on a geological map. The method is both neat and clear and much to be preferred to the tedious orthodox approach to such high values.

The cartographic rendering of the techniques presented few particular problems, so long as the maps needed only to be used to express trends and the same 'reference frame-work' of places was used as in Figure 47. Lack of different colours for negative and positive values was again a handicap and unfortunately with rather dissimilar ranges (the natural increase figures are almost all positive but the migration figures are balanced fairly evenly about zero) it did not prove entirely practicable to maintain comparable shading types for roughly similar values on both maps, shading as opposed to dot stipples beginning at $+9.5$ on the natural increase map but at -7.5 on the migration one; had two colours been available a greater variety of shading would have been possible and this small objection would have been quite easily overcome.

So much for the first approach. Let us leave an assessment of the results until those of the second method, with its more intricate cartographic problems, are available for comparison. The main problem involved in the second approach is to come to a sufficient understanding of the figures before we begin the cartography. A thorough analysis of both totals and components is needed and fortunately we have already encountered, in the scatter graph, a simple statistical technique which can be used to analyse the different ways in which two components occur together. In Chapter 2 the scatter graph was used as a means of studying relationships between two components which were rather separate entities (e.g. houses and people), but in this case the two components not only both refer to the same thing (people) but are expressed in a uniform fashion, using values which are capable of being added or subtracted from one another to produce an end product which is also under investigation. This peculiar combination of circum-stances enables much more information to be obtained from the scatter graph than is normally the case and a little 'thinking aloud' about what happens on the graph in this case will help understanding of the distribution and analysis of the final figures.

Figure 48 illustrates some of this thinking aloud. Consider, for example, what happens if we plot on the graph a series of points whose individual figures are as follows (natural increase first and all values in percentages):

A −5, 5; B 10, −10; C 20, −20; D −5, 10; E 10, −5; F 20, −15; G −5, 0;
H 5, −10; J 15, −20.

The nine points consist, in effect, of three sets of three, with total changes of 0, 5 and −5 respectively and it will be seen that each set lies along a straight line at an angle of 45° to each axis. A little thought will show that any number of similar diagonal lines could be drawn on the graph and that all points along any one such line would have the *same total change*, irrespective of the size of the components; conversely, we can use lines like these to divide up the scatter of points on the graph and produce groups which have total changes above or below the particular value selected. It is not difficult to draw in the line representing any particular total change. All that is needed is to devise two pairs of components which will produce this required change, plot them on the graph and join the plotted points by a straight line.

Thinking about the graph in this way soon enables one to appreciate further significant ideas, such as the implication of a point being situated in different parts of the graph. The line ABC for example divides the graph into two parts, with total change positive above that line and negative below it. Similarly, along certain lines specific things can be said to happen. Along XX, YY and ZZ, for example, one factor is always zero, the other factor therefore being entirely responsible for any change, and from this it is an easy step to realising that for points *close* to these lines one factor will be very small, probably negligible. We can also find other lines such as PP where both factors are equal, but are accentuating one another because both are positive, or a line like QQ where one factor is twice the other. From lines such as these it is not difficult to label the intervening areas of the graph as having certain characteristics which will help to subdivide sensibly the scatter of plotted values when these have to be studied.

The actual scatter graph used in the analysis of the figures is reproduced in Figure 49. At first sight the points appear to lie awkwardly widespread, but a second glance shows that there is perhaps a crude tendency for many of them to lie along a broad diagonal band astride the position of the 'no change' line (see Figure 48)—in other words in many of the areas total changes are small, chiefly because negative migration is counteracting positive natural increase. This is a useful beginning to a classification of the results and it would seem sensible to try to introduce into it some idea of the amount of total change. A straightforward *dispersal* graph was made for the total change figures and indicated that values of plus and minus 1·2% and plus and minus 6% would act as suitable demarcating points, and lines for these amounts of total change were therefore plotted on the scatter graph. Unfortunately, the three categories so defined contain places of similar total change, but very differently sized components, some being due to the interaction of values of less than 1% whilst others have values both more than 20%. Each of the three basic divisions was, therefore, further subdivided into three according to the *size of the components*, simply by drawing horizontal and vertical lines across the bands. So that a reasonably concise 'label' could be added to each group it was desirable that the limits for both migration and natural increase should have the same values, and in this case the limits of $\pm 7.5\%$ and $\pm 15\%$ divided each band conveniently into three sections. With these limits only four points fall in the stippled areas on Figure 49; they comply with one but not the other limiting value, but the use of 2% tolerance limits allows them to be included in the most suitable adjacent group.

About half the plotted points in Figure 49 fell within the classification so far devised and the remaining values fortunately presented rather clear characteristics, corresponding roughly to three groups situated in parts of the graph which indicated that migration, natural increase and negative migration were dominant factors respectively, and a fourth group in which the most unusual feature, if not the largest, was a *negative* natural increase. Each of these four groups could be further subdivided according to whether total changes involved were in excess of 16% and 26% if positive and -12.5% if negative, all of these additional limiting values being suggested by a dispersal graph of total change for all plotted points. The total effect produced by all the various limits imposed on Figure 49 was to divide the results into eight groups, based on amount of total change, subdivided in each case into two or three categories according to the dominant factors or nature of the main components.

The representation of this twofold classification presented something of a notational problem which was solved by devising eight basic types of shading for total change, supplemented by letters to indicate the componental aspects. 'O' in three sizes was invoked to show those cases where figures tended to cancel each other out and 'M' or 'N' to indicate areas where migration or natural increase were dominant; no sign was attached to the 'M' or 'N' as this was presupposed from the negative or positive nature of the basic shading.

The application of this technique is illustrated in Figure 47B (p. 147) and the result is not unlike Figure 47A in many respects. This is hardly surprising, for the technique, a straightforward shading map, is basically the same, but in this case the usefulness of the specific information on *amount* of change is enormously increased by the additional information concerning its *nature* or more especially its dominant component. There are also certain detailed points of difference between the two maps; only eight types of shading have been used in Figure 47B to fit in better with the more complex considerations of the dual notation and from this point of view the necessity of carrying two aspects of change has produced a slightly less specific map. Unlike the previous example negative and positive values have been differentiated by different types of shading (dots and lines respectively) with the blanks being retained for values around zero. These near-zero values almost always coincided with areas where components were cancelling one another out and the blank space seemed particularly symbolic of this 'not much-happening-in-the-end' state of affairs.

Comparative assessment of results and conclusions
Through the medium of Figures 46 and 47 the relative merits of the two approaches to this particular problem can now be compared. If the first or orthodox line of approach is

taken it cannot be doubted that Figure 47A gives a clear picture of total change and allows an easy appreciation of the main points regarding this. In the left-hand portion of the map outside the larger, named towns total change is usually either small or slightly negative, although the named towns themselves generally show rather larger positive changes, and a well-defined area of marked increase occurs around or adjacent to many of them, especially Leeds and Bradford. In the western corner of the map towards Todmorden, in an area of very marked relief, quite large negative total changes occur.

In contrast to this relatively simple picture the right-hand half of the map exhibits much more complicated patterns. The southern area around Barnsley and Rotherham is one of general increase, often quite large in amount though very patchy in detail, and with suggestions of the same broken rings of greater increase around towns such as Rotherham as occurred around the towns in the west. The eastern corner of this section of the map, however, is generally an area of marked decrease with the obvious exceptions of a belt east of and including Pontefract and a 'pepper-pot' distribution of isolated high values which is typical of the whole of this right-hand portion of the map.

Despite these easily drawn conclusions it must be remembered that a simple realisation of the amount of total change is not the main object of this particular exercise and the question which must now be answered is how far we can effectively appreciate the differing backgrounds of these types of total change by a perusal of Figures 46A and B. Fortunately, the patterns indicated on Figure 46A are simple and relatively easily committed to memory, being composed simply of a steadily increasing trend in a clockwise direction from large negative values in the western corner to a maximum along a line almost due east from Barnsley; a trough of rather lower values occurs between that town, Rotherham and Sheffield, but there is renewed increase east of Rotherham again. Figure 46B is another matter, however. Unlike natural increase, migration shows a less rational pattern and many of the complications of Figure 47A appear to have their roots in the intricate patterns found on Figure 46B. As in Figure 46A the left-hand portion of the map is relatively simple with areas of marked 'suburban increase' around and between Leeds, Bradford and Halifax, but in the right-hand portion the extremely complicated patterns almost defy general description apart from the Pontefract 'ridge' and the area of marked decrease around Doncaster.

The unfortunate coincidence of the most intricate portion of Figure 46B with the only portion of Figure 46A where patterns become at all complicated, points to the inevitable conclusion that only a person remarkably well blessed with visual perception and awareness will be able to supplement Figure 47A with anything other than the most vague and general observations compounded from Figure 46.

By comparison with this hit-or-miss method the much more elaborate Figure 47B, which illustrates the second approach, seems to guarantee more positive results. Admittedly, Figure 47B is more complicated than Figure 47A, so that it is a map to be inspected and sampled rather than glanced at, but inspection will produce clearly and incontrovertibly the answer which is being sought. It is now possible, for example, to differentiate easily between the discontinuous 'rings' around Leeds and Bradford where migration is dominant and that around Barnsley where the retention of a high rate of natural increase is playing a much larger part. Contrast too the areas north of Otley and around Todmorden; both show considerable decreases, but in one emigration, in the other negative natural increase is the main cause. In like fashion the area of small change or light decrease which runs horizontally across the middle portion of the map is generally produced by opposing, cancelling factors of small magnitude around Bradford and Halifax, but of much greater size between Wakefield and Barnsley.

Such a comparison of the effectiveness of the two approaches underlines the observation already made in Chapter 4. The practice, still commonly encountered in articles, books and national atlases, of presenting related information in the form of a series of maps in the hope that these can be assimilated in compound manner by the map user falls far short of what is really required. The responsibility of the process of fusion lies with the cartographer not the observer, and demands more subtle and sophisticated approaches if it is to be successful. The *apparent* complexity of the end-products of this second approach, disconcerting though it may appear at first glance, is much to be preferred to the inefficient and misleading simplicity of more orthodox solutions.

Example 4   A distribution concerned with multiple variables. Agricultural land use in Sweden, 1959 (Figures 50 and 51)

The examples so far described have all been concerned with situations in which no more than one or two variable factors had to be considered. Not infrequently, however, distributions with multiple variables occur and have to be portrayed visually, the most commonly encountered examples of this type of distribution being found in statistics relating to land use or employment. In these cases the complete entity of total land used or numbers employed is presented as being divided up into several categories of different land use or types of industry, the proportions of these varying at each place for which figures are given. As an example of problems involving this type of distribution, an attempt is made here to illustrate the variations in agricultural land use throughout Sweden in 1959.

Source and aspects to be considered
The basic source of the figures used in this particular problem is Table 60 of the *Statistical Yearbook of Sweden* for 1960. In that table the total acreage of land in agricultural use is given for each of the twenty-four counties of Sweden and this total is further subdivided into land under seventeen major crops or uses, whilst Table 46 in the same source enables this agricultural area to be seen against the total area of other land uses (forest, unused land, built-up areas, etc.) in each county. The aim of this study is to illustrate this changing pattern of land use by means of a map which will produce a *fusion of impressions* illustrating not merely land use but the varying pattern of agriculture in Sweden, distinguishing changes in emphasis from one part of the country to another and picking out the marked similarities which occur between region and region no less than their differences.

General statistical considerations
Any comparative study of figures relating to different administrative areas demands statistics presented in comparable form and the first major task here is to translate the

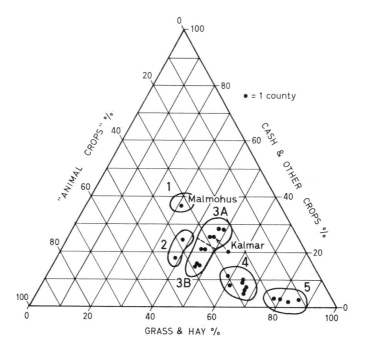

Figure 50 A triangular graph to show allocation of agricultural land in the counties of Sweden among three constituent uses.

Source: Statistiska Centralbyran, *Statistisk Arsbok för Sverige, 1960*, Stockholm, 1960.

values for the actual number of hectares in the seventeen use classes in each of the twenty-four counties into percentages of total land in agricultural use. At this point, however, anticipation of the later stages of the work suggests that some of the 408 calculations involved might be superfluous, for the simple fact is that *we cannot deal with seventeen variables.* Admittedly, two methods described in Chapter 2 (divided circles and divided rectangles) were adapted to illustrate variable components and could certainly accommodate seventeen sectors or slices, but the main problem here is mental and visual rather than mechanical. If we had misgivings about the simple comparisons involved in the previous example it will be readily apparent that we shall not get far with a visual assessment of the changes in twenty-four examples with seventeen variables in each. *The first requirement, therefore, is to reduce the number of variables to more manageable proportions.* In effect this has been achieved by reducing their number to *three* and several advantages accrue from such a drastic telescoping of the detailed picture. In the first place we have already encountered a device, the triangular graph, which will enable us to plot entities composed of three (but no more) components and allow the easy recognition of similar and dissimilar occurrences, so that there is a strong, statistical reason for choosing such a number. Fortunately, agricultural considerations do not run counter to this. Despite its many detailed variations, Swedish agriculture (the same is true of many countries) can be

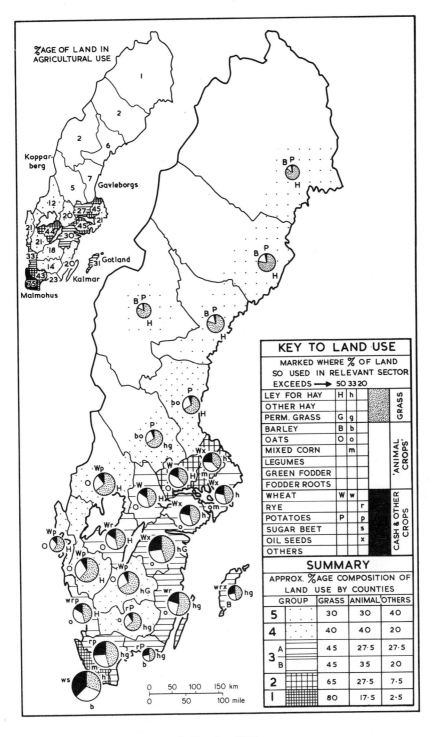

Figure 51 Agricultural land use in Sweden, 1959.

Source: As Figure 46.

viewed as a dual activity—namely, the raising of livestock with all this implies and the growing of crops for other purposes, either for human consumption locally or for sale. So far as the livestock sector of the economy is concerned there is in Sweden a dual basis of grass or hay from either permanent or rotation grassland, and the growing of crops for animal feeding-stuffs so that a threefold division into grass and hay (grass), crops grown for animal feeding-stuffs (animal crops) and crops grown mainly for human consumption or sale (cash and other crops) will effectively delineate the main pre-occupations of the Swedish farmer. This division is not entirely rigid of course. Some crops such as hay, fodder beet and sugar beet fall clearly into one of the main categories, but others such as barley or potatoes could find a place in two. Where this occurs these crops have been consistently included in that division which accounts for the bulk of the acreage—thus barley is always considered as an 'animal crop', potatoes as a 'cash crop' because this is predominantly the case throughout most of Sweden.

With such a threefold division it now appears necessary only to calculate the total in each of the *groups*[1] as a percentage of the whole and many detailed calculations have been eliminated. Of necessity the figures for 'fallow and unused arable land' have had to be eliminated from the total; they could not be allotted to any of the major groups and to have introduced a fourth sector would have destroyed the whole basis of the subsequent analysis. Since land in this category usually occupied between 3% and 7% of the total agricultural land the omission, though awkward, is not serious.

Choice of methods and details

Having reduced our original figures to manageable proportions, we are now in a position to elucidate the broad patterns which these suggest by plotting the percentage figures for each of the three main sectors in each county on a triangular graph as in Figure 50. The results there indicate that despite their variety the figures for the twenty-four counties can be simplified into five groups with varying degrees of precision. Some groups, e.g. groups 4 and 5 are clearly defined, others, e.g. groups 2 and 3, much less obviously so, and group 1 which contains the solitary country of Malmohus emphasises the unique position of that remarkable county in Swedish agriculture. Only one county, Kalmar, lies awkwardly outside any of these five groupings and since it is by no means as important and unique as Malmohus it has had to be included 'tolerantly' with group 3A with which it shows most similarities. In this case the resultant groupings have been defined in Figure 51 in terms of the average values for each of the three components, an easier but less satisfactory method than describing the groups verbally. The quoting of upper and lower limits only, to define groups, has already been condemned since the six figures per group which it produces are too numerous to be assimilated easily. It is possible however that when groups are conveniently compact and values can be rounded off (both are the case here) that quoting typical or average values might be considered as a substitute for verbal description.

As Figure 51 shows, this is a most useful preliminary as the juxtaposition of the counties within each group does in fact divide Sweden up into some half-dozen or so more or less continuous areas, each with a characteristic basic agricultural pattern to which the detailed information can now be added.

[1] For the crops included in each group see the key in Figure 51 (p. 157).

Although seventeen initial variables have been reduced to three it was never intended that the original seventeen should disappear completely from the picture. Some form of representation of the distribution of the more important of these crops is just as necessary as representation of the overall pattern, but it is essential that this latter feature should be established first so that the detail is seen against an appreciable background and not simply as profusion of isolated occurrences.

Fortunately, the use of a shading technique for the first, general considerations will permit the inclusion of another 'single-symbol' technique superimposed on the shading, and for this purpose the threefold subdivision was represented by means of a pie graph for each county. Each of these pie-graph symbols enables the actual variations in the component parts to be shown and also acts as a base against which detailed information concerning specific crops can be placed. As with earlier considerations, the temptation to add a profusion of detail must be avoided and it was, therefore, decided to indicate only the really important crops which contributed to each sector, these being easily added by placing a series of letters against the relevant section of the pie graph. Crops considered sufficiently important to warrant particular mention were those which occupied 50% or more, or 33% to 50% of the land *within their particular sector*. In the case of some cash and other crops where the greater variety possible meant that such figures were rarely encountered the figure of more than 20% was substituted as a lower limit (see Figure 51). The use of such simple figures for subdivision meant that the need to include or exclude any crop from the notation could almost always be determined by observation without calculation, and the decision to base the percentages on land *within a particular sector* and not on percentages of *total* land meant that changing regional emphasis within the broad framework already defined would be well illustrated.

Further points of detail are the confining of the background shading in the northern provinces only to those areas where agriculture assumed any importance, and the addition of an inset map showing the percentage of total land in agricultural use, to indicate how widespread or otherwise were the conditions represented on the main map.

General conclusions derived from Figure 51

Figure 51 emphasises once again the success of this compound method over the more orthodox approach of multiple-sectored unannotated pie graphs so often used for this type of statistics. Not only are valid basic patterns of Swedish agriculture established from out of a mass of detail but similarities in detail are also emphasised and the distribution of individual crops and transitional tendencies in the crops grown in each sector are well illustrated. Consider, for example, the way in which the map brings out the transitional tendencies in the animal crop sector from the extreme dominance of the very tolerant six-rowed barley (B) in the north, though a gradually increasing acreage of oats (b.o.) in Gavleborgs and Kopparberg, to the extreme dominance of oats (O) further south and the re-emergence of barley (this time of the two-rowed variety) in the extreme south and in Gotland. Regional specialisations such as potatoes (P) or (p) on the sandy soils of the south-east, or oil-seeds (x) in the eastern counties emerge quite clearly, whilst consistent regional associations in all sectors, e.g. BPH in the north, WHO, WpHO and Wxhm in Central Sweden are all apparent.

8 Computer mapping

The techniques which have so far been described in this book have envisaged as their end product a hand-drawn statistical map or diagram. At its best such a product has many advantages—boldness, colourfulness, and clarity for example—but almost inevitably these are aquired as a result of slow, laborious, and hence often very expensive, preparation. Not surprisingly therefore cartographers have sought ways and means of mechanising and speeding up this process and to this end the last ten years or so have seen very rapid, varied and exciting developments arising from the employment of computers and computer-driven devices to prepare statistical maps.[1] Of course a major defect of computer produced maps is that their preparation is restricted to the very limited number of people who have access to such specialised equipment, but while hand-drawn statistical maps must still remain the norm for the vast majority of cartographers the computer produced product is likely to become much more common and it is essential that all those dealing with statistical maps have at least some awareness of actual and potential developments in this field. It is the aim of this chapter to produce some such awareness; it is not, and cannot be, a detailed manual for the techniques involved but seeks instead to provide an understanding of the principles which underlie them and the range of 'computer maps' which can be produced.

As the illustrations to this chapter indicate this range of computer produced statistical maps is surprisingly varied, yet its variety springs from the use of only two major basic groups of techniques, which in turn result from the employment of two different basic types of equipment. These are *line printers* and *graph plotters* and the remainder of this chapter considers in turn the characteristics of maps produced by these means.

8.1 Statistical maps produced by the use of line printers

To date by far the most commonly encountered types of computer map are those produced by line printers, simply because the line printer is an integral part of the output side of any computer and can therefore be used to print maps without the need for any additional specialised pieces of equipment.

The line printer itself is the part of the computer which prints the output in readable

---

[1] This adaptation has not been confined to producing statistical maps. The preparation of block diagrams, perspective drawings and some aspects of topographical maps can be 'computerised' in similar ways. The very varied end-products are sometimes referred to under the collective title of 'computer graphics'.

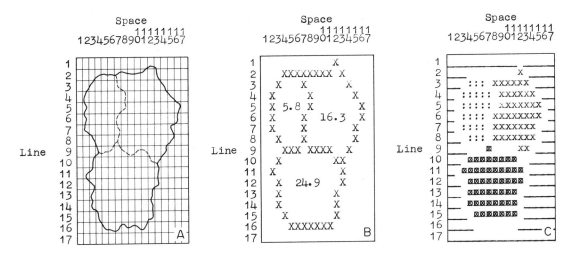

Figure 52 Producing maps by typewriter. Similar 'print patterns' can be produced mechanically by the line-printer of a computer. **A**. A map of a hypothetical island divided into three parishes. **B**. A type-printed approximation to this map with information about the population density of each parish added. **C**. A type-printed shading map of the distribution in **B** (no key is provided as the example is for illustration of shading only).

form and is best envisaged by the layman as a kind of typewriter whose keys are activated by instructions from a computer instead of the conventional typist.[1] How can such a machine print maps? The answer is quite simple once one basic problem is overcome. All features on a map occupy *positions in space;* if these positions can be identified in such a way that a 'typewriter' can produce them in terms of line so-and-so, column so-and-so an approximation to the map can be produced. Consider for example a map of a small island divided into 3 parishes as in Figure 52A and occupying an area 17 lines high and 17 spaces wide on a conventional typewriter. If we observe in which spaces on which lines particular features such as the coastline and parish boundaries occur (ignoring those spaces where the line merely cuts across a corner) and 'tell' the typewriter to print an X in these particular spaces we shall produce an approximation to the coast line and boundaries as in Figure 52B. With a computer the same process can be performed by feeding into it the necessary information (e.g. line 1 print X in space 12; line 2 print Xs in spaces 4 to 11, 13; line 3 print Xs in spaces 3, 7, 14 and so on), translated onto punched cards or magnetic tape; the computer then 'instructs' the line printer and the map appears.

This is only a beginning however. The computer can also be provided with specific information about each area, e.g. the population density per square kilometre of each

---

[1] Other important differences are that the line printer prints a whole line at a time at speeds greatly in excess of normal typing speed onto paper of restricted width (about 330 mm composed of 130 characters usually) but almost limitless length. It has about 150 characters including one set of letters (capitals) and many symbols. Additionally special 'print trains' can be obtained containing more characters (lower case letters for example) and even specially designed mapping symbols. Should an individual map require a greater width than the normal 330 mm the computer can calculate and produce it in the form of two or more strips, which can then be joined together.

parish in our hypothetical island, and can be told to print this value starting at a pre-selected and convenient space (line 5, space 4 for parish A, line 6, space 10 for area B etc.) at the same time as it prints the outline and boundaries, as in Figure 52B. If the computer is given a data bank of say 200 different kinds of information about each parish it can be instructed to take each item in turn and print a map of its distribution in the manner described above so that 200 maps each dealing with one topic in the area can very rapidly be produced. Equally simple is the idea of abandoning printed values and replacing them by a form of shading. The range of values involved is examined, divided up into suitable groups and a 'printable' set or 'bank' of shadings devised to represent these groups. By examining each character-space[1] within the area of the map, identifying in which parish it occurs, referring to the value for that parish in its data bank and the type of shading appropriate to that value as defined in its 'shading bank', the line printer can be instructed to print the necessary symbols in each character-space and so build up a complete shading map of the area.

The only new problem introduced with this idea is that of devising a range of 'printable' shadings, but this is not difficult to do for in addition to printing single characters the line printer can over-print as well. By using letters (or symbols) and overprinted letters it is possible to build up a range of 7 or 8 symbols with distinctive tones of greyness. Boundaries can be emphasised by omission (i.e. leaving a blank space where they occur), as in Figure 52C, or by having the blackest symbol of all to mark them; areas beyond the map may be left blank or distinctively shaded as in Figure 52C.

The relatively simple principles underlying the process just described should not be allowed to disguise the very substantial amount of data preparation which they require. Consider for example the problem of defining the boundaries of each parish and hence the character-spaces which lie within it. This could be done, as in Figure 52A, by preparing a base map of the area to the scale of the eventual line printed map, covering it with a grid of rectangles each corresponding to one character-space and then systematically identifying the position of each boundary and any other required position, e.g. the locations at which to commence printing information as in Figure 52B. For a large map with many boundaries this would be a lengthy and tedious operation and fortunately much of the work can be mechanised using a *digitiser* together with the calculating capacity of the computer. The digitiser is best envisaged by the layman as a sort of mechanised grid-reference reader.[2]

Placed over any position on a map, e.g. one of the corners of the area to be mapped, the digitiser will record the co-ordinates (grid reference) of that position on tape in a form readable by a computer; moved along a boundary line stopping at each important change of direction it will record the co-ordinates of such points whose location will provide an approximate definition of the boundary. If this recorded information together with some basic decision about the overall size or scale of the finished product is then fed into the computer the machine itself will identify the character-spaces in which the

---

[1] The term character-space is used here and subsequently to define any unit which can be identified as 'line so-and-so, space so-and-so' and in which the line printer can therefore be instructed to print a character.
[2] In practice actual grid references or an independent system of coordinates which work very similarly to grid references may be used.

boundary lines, etc., occur without need for further manual intervention. An obvious by-product of this process is that if for any reason a change of scale in the resultant map is desired the computer can easily prepare a larger or smaller version simply by re-calculating positions. In addition the computer can also calculate the areas of the units defined by these boundary lines, as well as the positions of their centroids.

Once the obvious step of utilising the calculating (as opposed to merely storage and printing) capacity of the computer has been taken it is easy to see how the range of techniques can be enormously enlarged and even more tedium removed from preparation. Returning to the hypothetical data bank of 200 different types of information on parishes the computer can make far more than 200 maps simply by combining aspects of one element with another. It could produce, for example, maps showing the ratio of item 1 to item 2, the percentage which item 3 forms of the total of items 3, 4, and 5, and so on. As for labour saving on the design side there is no need, for example, to make a previous examination of data to determine the numbers of shading groups and their limits. The computer can be asked to scan the range of values involved, divide it up on a preselected plan, and then select the required number of shading symbols from a bank of shading symbols which have been fed into it. The method of division into groups can be in terms of equal intervals, quantiles, standard deviation units (see p. 83: the computer will of course calculate the mean and standard deviation itself), or any other system of sub-division which can be described in a set of specific instructions;[1] any one of these methods could be used or the computer could be programmed to produce several maps using 2 or 3 different methods so that the most attractive and suitable could be selected—an almost unknown luxury where hand drawn statistical maps are concerned.

By these means the production of almost any sort of shading map could be computerised and indeed new types impossible to produce by hand can be added. A major weakness of shading maps is that the shapes of shading patterns found on them are determined by the vagaries of boundary lines which are themselves often totally irrelevant to the feature being mapped (see p. 56), a defect which can be overcome using a computer-prepared *proximal shading map*.[2] In this case the value for a particular area is envisaged as a 'spot-height' (or spot value) located at a suitable character-space at the centre of that unit. The computer then scans each character-space on the map, calculates which spot value is the *nearest* to it and then prints in the space *the shading symbol corresponding to the nearest spot value* (hence the term proximal). In this case boundaries as such are removed from the map, together with the need to feed into the computer the lengthy description of their position; all that need be given is the co-ordinate locations of a few spot values instead.

Once the idea of 'spot-heights' is introduced it is natural to think of those statistical maps which use isolines and in particular to consider the idea of using the computer to calculate the interpolation and positions of the isolines. Several computer-mapping

---

[1] It would be impossible, for example, to request the computer to find 'natural breaks' (see p. 87), unless some specific method of defining these could be devised.

[2] Also known as a Thiesson Polygon map. The author confesses to preferring the descriptive term 'proximal shading map', the word 'proximal' having been first applied in this context by Professor H. T. Fisher of Harvard University.

techniques, are available to carry out this procedure, one of the more flexible and best known being that included in the SYMAP programme.[1] Essentially what the computer does is to calculate the value or 'height' of each character-space by interpolating from at least *four* adjacent spot values; it can then print in that space the shading symbol appropriate to that value, as in Figure 53, or alternatively can find the character-space which most nearly coincides with the actual isoline value and print some symbol there, leaving the rest of the map blank so that the isolines appear in the same manner as the boundaries in Figure 52B. Either way, of course, the process has an accuracy which could not possibly be matched by the simple estimate-interpolation used in the hand method (see p. 58) though it is only fair to point out that the sophisticated calculations involved are also beyond the capacity of some of the smaller computers.

Shading and isoline mapping techniques can therefore fairly easily be 'computerised' using line printed output. The dot map, as might be expected, produces rather more intractable problems but an interesting line-printed compromise has emerged in the LINMAP system devised by the Planning Services section of the British Department of the Environment. The basis of the 'dot map' in the LINMAP system is again spot-values, preferably though not necessarily, referring to relatively small, closely spaced areas— wards in towns and parishes in rural areas have been used for example. For each areal unit the co-ordinates of the centre of its built-up area are determined and these together with statistical information about that unit are fed into the computer[2] which translates the co-ordinates into the nearest character-space equivalent and prints there a *single* symbol corresponding to the value of the topic under study. The essential difference from the procedures described above is that *only one symbol* is printed for each unit and character-spaces where no spot value occurs are left blank; alternatively should two or more spot values occur in the area occupied by any one character-space their values are amalgamated or averaged as may be most appropriate.

An obvious weakness of such a technique is that it might result in a map in which a few small point-value symbols 'float' in a great blank background, totally unable to produce any visual impression of overall value distribution. This is avoided however by preparing the finished maps at relatively small scales. In this way the actual ground area corresponding to each character-space on the map is much enlarged and many or most character-spaces will therefore contain at least one spot value[3] so that far less of the map is occupied by blanks, which may indeed disappear completely in areas where spot values are closely spaced. A typical result is shown in Figure 54, though here the spaces which should be blank have in fact had a 'filling symbol' of a full stop, printed in them.

[1] SYMAP-Synagraphic Mapping System. Many standardised computer-mapping systems and techniques have distinctive 'trade names'; the SYMAP programme devised by H. T. Fisher at Harvard University is one. It contains programmes for producing proximal and shading maps as well as isoline maps; other such programmes are referred to later in the text. (It may be noted in passing that when applied to a computer the word 'programme' is often spelt in the American manner.)

[2] E.g. for dealing with material from the 1966 census a data bank of up to 381 different items ('up to' since not all items were available for all units) for each area was prepared. Maps may be made from any of these items or from certain combinations of them.

[3] They may in fact, if data areas are small and the scale of the map is small too, contain several, which have to be 'amalgamated' and represented as one figure.

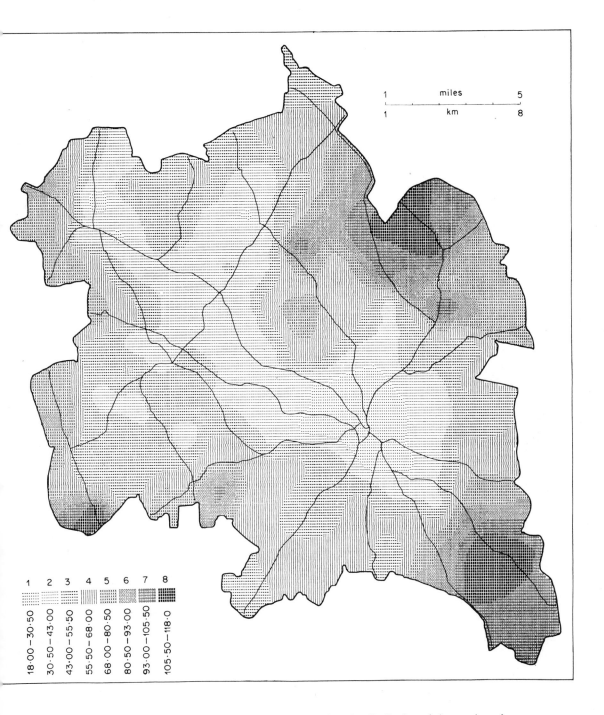

Figure 53 An isoline map produced by line printer and showing the distribution of the number of cars per 100 households in the West Midlands Conurbation, 1966. The basic values were derived from ward data from the 1966 Census: the eight shadings are produced from simple letters or symbols or combinations of these (groups 7 and 8).

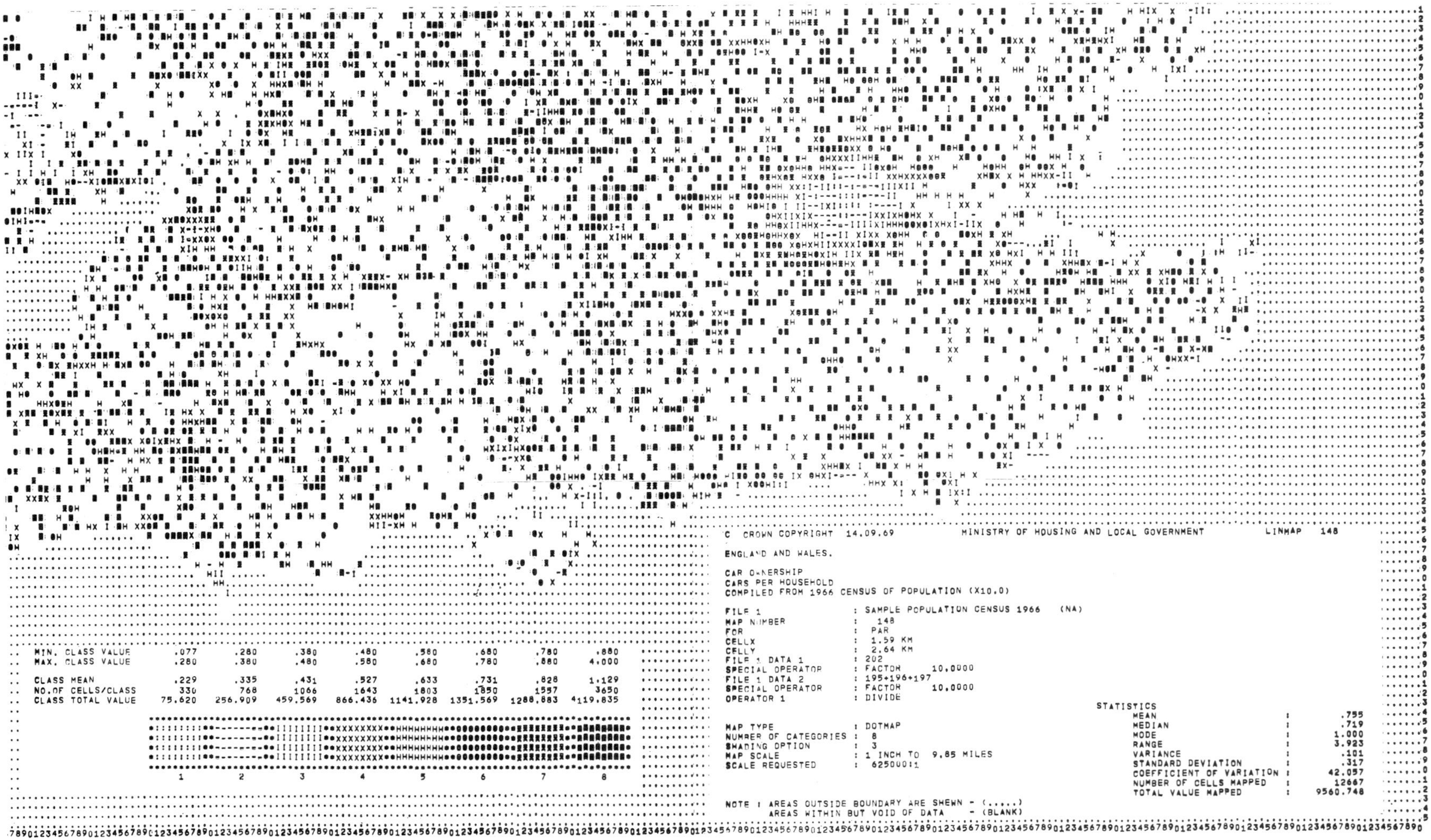

Figure 54 A Dotmap in the Linmap series. It shows the spot-values of a particular variable (cars per household) which occur in any given rectangle 1·59 km × 2·64 km (the size of one character space at this scale). If no value occurs a blank is left. Notice from the right hand key that the computer has calculated a good deal of information about this variable as well as mapping it.

Source: Department of the Environment.

This does not represent a value; its purpose is merely the better to differentiate the area of the map from the surrounding spaces. In other versions true blanks are left within the map and the surrounding space not covered by it is filled with full-stops. As can be seen in Figure 54, the LINMAP 'dot map' is really an unusual kind in which each value is represented by only one dot which can itself vary in size and intensity.

Also contained within the LINMAP system are two types of shading map the 'zonemap' (a conventional shading map) and the 'gridmap'. In the latter case the area unit used for mapping is a grid square, usually a 5 km or 10 km grid square, which is represented on the map by a block of 15 character-spaces, 5 wide by 3 high.[1] In this case the computer identifies from their co-ordinates all spot values lying within a particular grid square, amalgamates these into one 'representative'[2] value for the whole square, and then prints the symbol corresponding to this value in all 15 character-spaces forming that square, producing a shading map in which all units have the same shape, (see Figure 55). As with 'dot maps' if any 10 km square contains no spot-value it appears blank.

A criticism often levelled at line-printed statistical maps is that their appearance is relatively unattractive and, though not always of overriding importance, e.g., where maps are needed for research rather than for publicity or publication, there is some justice in the observation. Even for publication, where a 'working map' rather than a 'public-eye-catching' one is needed, line-printed computer-maps may well suffice and indeed the first computer produced atlas has already been published.[3] No doubt there will soon be many more. Even so interesting possibilities for producing more colourful and attractive computer-maps have been opened up by the COLMAP variants of the LINMAP programme, whereby maps of the 'dotmap' and 'gridmap' types may be transformed into fully finished, entirely mechanically printed output in up to 10 colours. The essential difference between the COLMAP variant and the originals is that instead of using the computer's line printer to produce the graphic output the computer is linked to a special type-setting machine which prints only one symbol, a square, in any character-space or for any grid-square space which requires one. Because this square symbol fills the whole of one character-space or grid square space, adjacent symbols touch one another to produce a continuous mass, without the breaks between characters which are found in line-printed output. To enable the necessary coloured maps to be produced, however, the computer instructs the type-setter to print not one map within n shading groups, but n maps each showing the distribution of *only one* shading group. Each of these n maps is then used to prepare a printing plate whose information is printed on the final map in a distinctive colour; when the information from all printing plates has been printed on to the map the final output in n colours is complete.[4] As with normal 'dot maps' and 'grid maps' blanks can occur,

[1] Because the character-spaces are rectangular a square shape can only be built up by balancing the width of 5 spaces against the height of 3 lines, see Figure 55.

[2] Just how 'representative' or 'real' such a value may be will depend largely on the number of and variations in the spot-values which are included. As with all spot-values standing for areas, the bigger the area the more difficult is it to let one figure represent conditions over the whole of it.

[3] K. E. Rosing and P. A. Wood, *Character of a Conurbation: A Computer Atlas of Birmingham and the Black Country* (London, 1971).

[4] The description given here is only one of three colour printing techniques possible with COLMAP. In fact the other options which use overprinting of colours to obtain many different shades are perhaps more likely to be used, but since the principles involved in the one colour = one shading group = one printing plate option are more readily described and understood by the layman, this method has been selected here.

PERCENTAGE OF TOTAL HOUSEHOLDS

23   36   49   62

KILOMETRES
10 0   80
10 0   60
MILES

Figure 55 (*above*)   Part of a gridmap in the LINMAP system showing the proportion of households in England and Wales owning their own housing accommodation in 1966. The statistical bases of the map are the spot-values which occur within any 10 kilometre grid square. If no value occurs in any square such a square is left 'blank', in this case filled with full stop symbols (e.g. Scottish Border area). The general impression obtained from the map is much improved if it is viewed from a distance of about 4 ft.
Source: Department of the Environment.

Figure 56 (*opposite*)   A COLMAP produced by the Department of the Environment showing persons per sq. mile in 1966. The statistical bases of the map are the spot values which occur in any 5 kilometre grid square. If no value occurs in any such square the square is left 'blank' that is in this case filled in in yellow, and this is the implication of the word 'area' in the key.

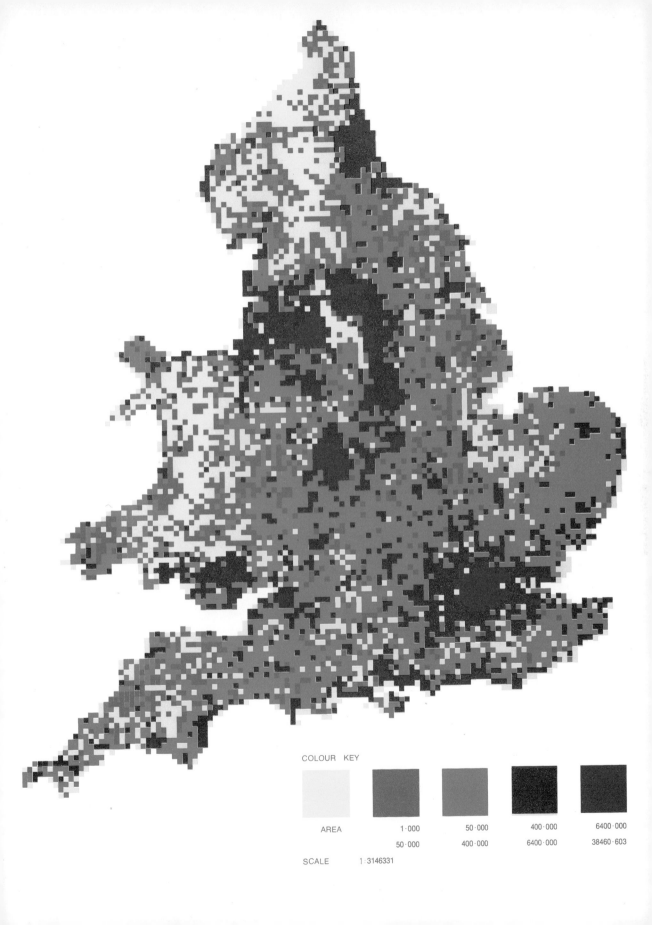

COLOUR KEY

AREA

	1·000	50·000	400·000	6400·000
	50·000	400·000	6400·000	38460·603

SCALE   1 : 3146331

but it is usually the case that these are printed in some pale background colour such as yellow, again so that the coloured map area can be clearly distinguished from the surrounding blank page. An example of a COLMAP is shown in Figure 56.

8.2   Statistical maps produced by graph plotters[1]

An inherent, though not necessarily serious, design limitation of maps produced by line-printers is that all lines and shapes have to be approximated using the relatively coarse unit of one character-space as the smallest dimension which can be shown. This is not the case in maps produced by graph-plotters which, as their name implies are instruments whose 'output-end' consists of a pen capable of drawing lines. The essential property of the graph-plotter is that when linked to a computer which can present to it the co-ordinates of a point, its pen will move to that point and there draw some dot, sign or symbol; alternatively given the co-ordinates of two points the graph-plotter will draw a line between them. Thus, provided that it is given sufficient co-ordinates to work from—say co-ordinates at every point where there is a major change of direction—it is capable of reproducing mechanically approximations to boundaries, coastlines etc., a property with obvious potential applications in topographical as well as statistical mapping. In effect no sloping line which the graph-plotter draws between points is ever straight, since its pen can only move either vertically or horizontally so that all movements must be reduced to a 'stepped approximation'. This is not a serious handicap however as the steps are extremely small, as can be seen in Figure 57, and the features which can be drawn at any point by the graph-plotter's pen are widely variable. It can be instructed for example to draw a circle of any selected radius centred at any point, or to write in symbols or names and values in either upright or italic lettering. The end-products of maps produced by graph-plotters are therefore of a rather different type and appearance to those produced by line-printers and Figure 57 shows 2 examples which have been produced by the MAPIT programme devised by R. Kern and G. Rushton of Michigan State University and the University of Iowa. In this programme the graph-plotter can be called upon to produce outline maps, dot maps (i.e. maps on which a single *drawn* symbol—a feature like an asterisk is often used, see Figure 57A—marks the point at which any feature occurs[2]), maps showing proportionate symbols such as a circle, and a new type of map not elsewhere described in this book, the *flow map*.

In this latter map all flows between one point and another are represented (irrespective of actual route followed), by a straight line joining the points concerned. Thus in Figure 57B which covers the city of Christchurch, New Zealand, the plotter using information supplied by a computer, has marked the position of each house interviewed in a particular study by an asterisk,[3] and each shopping and service centre by a circle whose area is

---

[1] Sometimes referred to as incremental plotters and drum plotters.

[2] This is not strictly a dot map but is very much akin to that in the LINMAP programme.

[3] Reduction from the scale of the original makes these asterisks appear as rectangular symbols in Figure 57B.

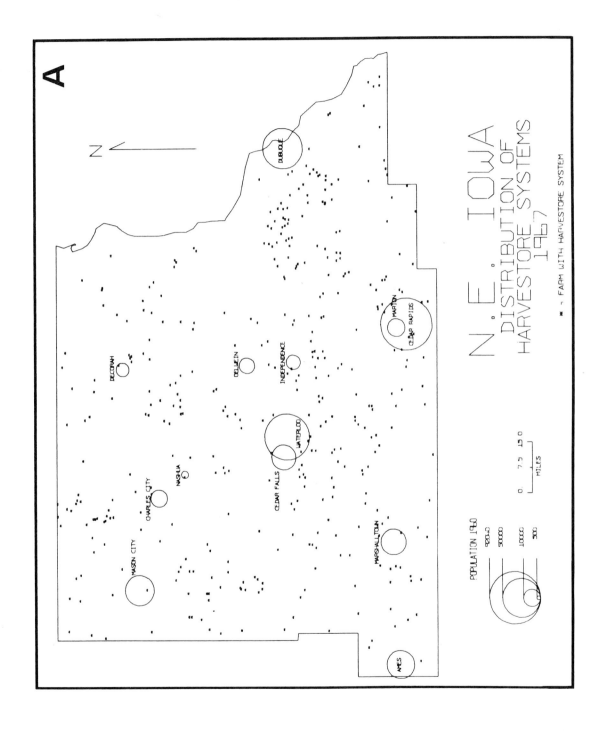

A

N

N.E. IOWA
DISTRIBUTION OF
HARVESTORE SYSTEMS
1967

▪ = FARM WITH HARVESTORE SYSTEM

DUBUQUE

DECORAH

OELWEIN

INDEPENDENCE

MARION
CEDAR RAPIDS

NASHUA

CHARLES CITY

WATERLOO

CEDAR FALLS

MASON CITY

MARSHALLTOWN

AMES

0   7.5   15 0
MILES

POPULATION 1960
200,000
50,000
10,000
500

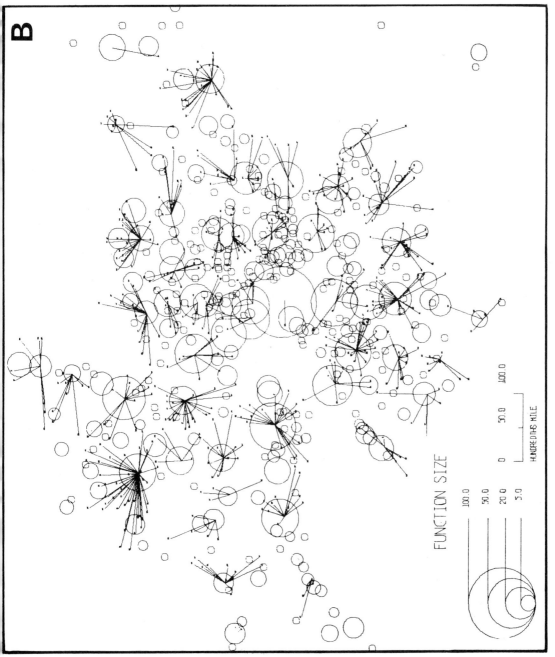

B

FUNCTION SIZE

100.0

50.0

20.0

5.0

0   50.0   100.0

HUNDREDTHS MILE

Figure 57 Two maps drawn by a computer with graph-plotter output using the MAPIT programme. In both cases the entire map including lettering has been plotted and drawn by the computer and graph plotter. **A**: Distribution of Harvestore Systems (asterisks) and populations of major centres (circles) Iowa, U.S.A. **B**: Locations of households interviewed (asterisks) and business centres (circles) in Christchurch, New Zealand. (See text for fuller explanation.)

Source: R. Kern and G. Rushton, 'Mapit: A Computer Program for Production of Flow Maps, Dot Maps and Graduated Symbol Maps', *The Cartographic Journal*, December 1969.

proportional to the number of functions found there. The information also indicated which centres provided a beauty care service and the plotter was instructed to draw a line from each house marked to the *nearest* beauty care centre. In this case the 'flows' mapped were theoretical—minimal possible flows so to speak—though in fact the plotter also used other information to produce another map joining each house marked to the beauty care centre actually patronised. Comparison of the two maps showed very important differences between the real and 'minimal' flow patterns.

8.3   Further developments of basic techniques

The discussion of techniques in the previous section was centred on the ability of the computer to produce maps comparable to those obtained using hand-methods. It must be stressed however that this was no more than a beginning in the field of computer mapping and that there are many modifications and developments of these basic techniques which make rather fuller use of the computer's potential in this field. An example of this, which has already been described, is the *proximal* shading map which, though simple enough in principle, needs the speed of the computer to perform the thousands of calculations required even on a small map. Other still quite simple developments, following on from methods which have been described above, occur when similar data for two different dates is provided; in this case the computer can print not only two series of maps, one for each date, but will calculate the changes which occurred between these dates and print maps showing these as well. It is already common practice to request that data be not only mapped directly but also mapped after additions, subtractions, calculations or comparisons have been made as well. For example if population data is supplied divided into say 20 age groups, maps can be printed by computer not only of the distribution of each age group but of any specific combination of groups, ratios of one group to another, percentages which these form of the whole and so on—calculations which would be feasible but undeniably tedious to perform manually. A similar idea is found in a computer map recently prepared to test the geocoding of the British 1971 Census data. Here for the County Borough of Huddersfield data on the number of households present was obtained for each 100 m grid square and mapped using a shading map with 6 symbol types. On the map which resulted and which was line-printed, each 100km square was represented by only one character-space, but instead of printing the value for the square in that space the value was added to those of the surrounding 8 squares and the symbol corresponding to the value of the *9-square average* printed instead. Furthermore, where no households occurred in a square, no calculations were carried out and the space remained blank on the finished map, which in its blanks therefore reproduced the patterns of the uninhabited portions of the borough and in its printed values showed household density by means of an moving spatial average which has the effect of damping down individual values and emphasising more general trends. Since in this case a rectangular character-space has been used to represent a *square* on the ground the relevant map appeared stretched slightly in a north-south direction (characters are taller than they are wide) but this is not serious and does not impede the usefulness of the map.

All of this, however, is still relatively simple manipulation of data; much more sophisticated feats can be performed. For instance if two sets of data are provided for a series of areal units the theoretical relationship between them can be calculated (by regression techniques), from this the value of one variable which *should* occur at any point, given a particular value of the other, is calculated, this value is compared with the real value and the *difference* (residual) calculated and then mapped—the whole performance being done entirely by computer and line-printer.

8.4 General observations on computer mapping

The types of sophisticated, data-manipulated maps described above, are indeed impressive and important, though it should be noted that it is the data which is mapped, rather than the type of map output, which marks them off from the typical manual product. Indeed, as will have been noticed, the mapping techniques are often the same— shading, isoline, and proportional circles being particularly common—and it is sensible to remember that even when drawn by computer many of the reservations which apply to these specific techniques are still valid. If a map with 72 different proportional circles is considered to be 'unassimilable' from the map-users point of view (see p. 76) it will *still* be so even though the 72 circles be drawn much more rapidly by computer and graph-plotter; if isoline patterns on many maps are meaningless simply because they have been interpolated from spot-values which are themselves meaningless and totally unrepresentative of actual conditions within an area, the isolines will still be of little use even though drawn with superhuman accuracy by computer. It is not impossible to imagine ludicrous maps where the accuracy of the interpolation contrasts glaringly with the inaccuracy of the spot-values on which it was based, and it must be stressed very strongly that *with computer maps as with hand-drawn ones the quality of the map can be no better than the quality of the data that is put into it.* There is here a very real possibility that simply because of the ability of computers to map large masses of data very rapidly the temptation may arise to map anything and everything without stopping to consider the quality of the input—or even the usefulness of the output. Alternatively, as we have seen, the very ability of the computer to perform lengthy and sophisticated manipulation and modification of data may in fact be the answer to the many hang-ups of data inadequacy which have rendered relatively useless so many hand-drawn statistical maps in the past. As has been stressed throughout this book there is far more to statistical cartography than simply 'turning figures into pictures', whether by hand or by machine, and it is in the knowledge and handling of his data quite as much as in the methods by which he ultimately presents it that the statistical cartographer's skill really lies.

9 Sources of statistics

In a work of this kind, where space is limited, it is always difficult to decide whether to include some description of sources of statistical information. The tremendous bulk and diversity of even the most obvious statistical sources has already prompted entire books devoted to this topic alone, so that any attempt to condense information into no more than one chapter seems scarcely worth making; yet even so two basic problems remain. The complete novice seeking statistical information for the first time needs guidance, even of the broadest possible sort, before he can get started at all; whilst the more experienced worker still finds unexpected difficulties which the elaborate existing reference works will not entirely help him to overcome quickly.

It is with the main aim of solving these two particular problems that this chapter is designed, and its arrangement follows from them. A very brief guide to the principal sources of statistics of an international kind is followed by a second section which is confined to describing sources of British statistics, also in outline form except in those sections which the author has found to present most difficulties in practice. This latter section in particular makes no pretence to completeness, and reference should also be made to the detailed works listed in the text.

One further general aspect remains to be mentioned. For the successful *mapping* of statistics *two* aspects must normally be considered—namely, 'How much or how many?' and 'Where?' Unfortunately, the same source does not often answer both questions and what may be described as 'quantitative' and 'locational' sources will usually be described separately and in that order.

9.1 Sources of statistics of a national and international nature

Perhaps the most easily accessible source of statistical information on the various countries of the world (though not generally world totals) is the *Statesman's Yearbook*, published annually in London. A typical volume provides, for each country, a 'hard core' of statistics on subjects such as area, population, administration, social services, industrial and agricultural production, imports and exports, and provides in many cases comparable figures for the preceding few years.

Similar but much fuller information, together with world totals, appears in the *Statistical Yearbook* published by the United Nations. The main headings under which information is provided are population, manpower, production (of all kinds), construction, energy, consumption, transport, communication, trade, balance of payments, international economic aid, wages and prices, national income, finance, housing, social statistics and education. Most tables show figures for the year of publication, the eight preceding years and one earlier year, currently 1953. A sister-volume, the *Demographic Yearbook*, adds more detailed information on population, births, deaths and marriages, and the *Monthly Bulletin of Statistics* also published by the United Nations provides figures on more specialised selected topics. A much simpler, but convenient, compact and cheap source of information on the more general aspects of world population, trade and resources (by countries) is *The Geographical Digest*, published in May of each year by George Philip, London. Climatic statistics, which are usually missing from the works already referred to, will be found in the *Tables of Temperature, Relative Humidity, and Precipitation for the World* prepared by the Meteorological Office and published in six parts by H.M.S.O., London, 1958.

Naturally, only a minimum of information about any country can be provided in a volume covering the whole world and for greater detail one must turn to works covering individual countries. Many states now publish a national yearbook, either descriptive with some statistics or, in many cases, entirely statistical in its nature. Examples of the latter type, taken mainly from Western Europe are *Annual Abstract of Statistics*, Central Statistical Office, published by H.M.S.O., London;[1] *Annuaire Statistique de la France*, Institut National de la Statistique et des Études Économiques, Paris; *Statistisches Jahrbuch für die Bundesrepublik Deutschland*, Statistisches Bundesamt, Wiesbaden; *Statistisches Jahrbuch der Deutschen Demokratischen Republik*, Staatliche Zentralverwaltung für Statistik, Berlin, and *The U.S. Book of Facts, Statistics and Information for* (year) which has now superceded the *Statistical Abstract of the United States*, both published by Bureau of the Census, Washington, D.C.[2]

Though differing in detail, these yearbooks have much in common, showing similarity of information, both among themselves and with the international sources just described. Thus the British 'Annual Abstract' has sections on Area and Climate, Population and Vital Statistics, Social Conditions, Education, Labour, Production, Distribution, Transport, External Trade, Finance, National Income and Expenditure, Banking and Insurance, and Prices. In these respects it resembles those of other countries, although in detail it is poorer on agriculture and land use, external trade by particular countries and cargoes handled at individual ports; detailed statistics for all of these are published elsewhere (see below).

Pertinent to the mapping of statistics is the fact that most yearbooks attempt some territorial breakdown of national totals, usually into those units forming the next tier in the administrative hierarchy (statistics for individual places are practically never given except for population, and trade at specific ports) producing figures for areas of greatly

---

[1] This can be supplemented by (1) a *Monthly Digest of Statistics*, (2) from 1953, a separate *Digest of Scottish Statistics*, (3) from 1954, a separate *Digest of Welsh Statistics*, (4) from 1965 an *Abstract of Regional Statistics* (see below). The Scottish volume is published by the Statistical Office, Edinburgh, the rest by H.M.S.O., London.

[2] Another very useful condensed source referring to the United States is the *County and City Data Book*, published about every five years by the Bureau of the Census.

differing size, some conveniently small, others so large that wide variations can still occur within their boundaries. These contrasts are illustrated in the following table where it will be seen that for British sources (the 'Annual Abstract') the smallest unit shown is the *country*. However since 1965 the annual *Abstract of Regional Statistics* (H.M.S.O., London) has provided totals for similar subjects for Scotland, Wales, and eight regional divisions of England, together with some *subregional* totals as well. No information is given by counties but the volumes contain useful definitions of the components of the main regional and subregional areas.

Country	Area 1.000s sq km	Administrative subdivision	No	Average Area 1.000s sq km
United Kingdom	244	Country	4	61·0
United States	9351	State	50	187·2
France	547	Department	95	5·8
W. Germany	248	Land	11	22.6
E. Germany	108	Bezirk	15	7·2
Netherlands	34	Province	11	3·1
Sweden	450	Country (Län)	24	19·8
Switzerland	41	Canton	25	1·6

In all cases only a minimum of information is broken down by areas, and national totals still remain of paramount interest. Most yearbooks attempt to provide comparable figures for past years, either in short runs or at selected dates and in some cases separate volumes have been published dealing solely with this historical aspect of national statistics. Examples are *Annuaire Statistique de la France, Retrospectif*, (I.N.S.E.E., Paris, 1961 and 1966); *Union Statistics for Fifty Years* (Bureau of Census and Statistics, Pretoria, 1960), and *Historical Statistics of the United States, Colonial Times to 1957* (Bureau of the Census, Washington, D.C.). The last-named volume in particular contains also a most useful text specifying sources of data, reliability, history of the statistics and much other allied information.

Unexpected windfalls in some yearbooks are an element of bilinguality (e.g. the Dutch and Swedish publications have table headings and footnotes in English) and an 'International Section' giving information similar to that in the *United Nations Statistical Yearbook*, though much reduced in volume. Especially useful in this respect are the French *Retrospectif* volumes which provide unexpected and convenient sources for historical aspects of major world production. Rather more detailed, welcome guides to whole jungles of official statistics are the two volumes *Statistics Europe—Sources for Market Research*, and *Statistics Africa* by Joan M. Harvey (Beckenham, 1968 and 1970 respectively), and which list, by countries the principal published sources of statistics for a variety of topics.

As we have seen, however, national yearbooks go only part of the way towards covering the 'where?' element of distribution and figures for even smaller areas must usually be obtained from national censuses. These are too complex to be elaborated in detail here, though many of them have strong similarities in content, published form and problems in use with the British Census described in the next section. It must be stressed, however that censuses are bulky multi-volume publications and in this respect a heaven-sent guide as to content and arrangement of all censuses, historical and recent, is provided by the five volume *International Population Census Bibliography* (Austin, Texas, 1965–6). Each volume covers a continent, as follows, No. 1 Latin America and the Caribbean (1965); No. 2 Africa (1965); No. 3 Oceania (1966); No. 4 North America (1966); No. 5 Asia (1966). Alternatively, an answer, complete or partial, may often be found in map form via one of the increasing number of atlases illustrating either world production or dealing with one particular country only. As with yearbooks, these atlases have much in common; contents are very similar and there is usually an accompanying text (including some actual statistical information) of varied length and quality. Unfortunately, actual statistics are rarely specified on the maps themselves and, as has been mentioned in Chapter 1, the impossibility in many cases of 'reading back' to the figures greatly reduces the usefulness of such atlases. Unlike yearbooks, national atlases are expensive publications, not widely stocked nor published annually so that they soon become considerably out of date. Details of national atlases which are in print today, together with a list of topographical and other maps arranged by countries will be found in *The Stanford Reference Catalogue* published by Edward Stanford, London, who also provide similar information for individual countries gratis.[1] The list below confines itself to ordinary commercially produced atlases which are more widely available.

N. Ginsburg, *Atlas of Economic Development* (Chicago, 1961).
G. Kish, *Economic Atlas of the Soviet Union* (Ann Arbor, 1960).
Oxford Economic Atlas of the World (4th Edn.) (Oxford, 1971).
The Shorter Oxford Economic Atlas of the World (3rd Edn.) (Oxford, 1965).
Oxford Regional Economic Atlas—The United States and Canada (Oxford, 1967).
Oxford Regional Economic Atlas—Africa (Oxford, 1965).
Oxford Regional Economic Atlas—The Middle East and North Africa (Oxford, 1960).
Oxford Regional Economic Atlas—U.S.S.R. and Eastern Europe (Oxford, 1956).

9.2   Sources of British statistics

The relatively brief description of sources of British statistics which follows must be examined in the light of the introduction already offered at the beginning of this chapter. Generally speaking the most pressing need of newcomers in this field is some guidance as to the general range and type of material which is available and this will be met by the Central Statistical Office's booklet *Lists of Principal Statistical Series*, published as booklet *No. 11* in the series *Studies on Official Statistics* (H.M.S.O., London, 1965), or by Appendix 1 of the Experimental Cartography Unit's *Automatic Cartography and Planning* (London, 1971). This last-mentioned work is a particularly useful and up-to-date list of the principal data sources in Britain (censuses of population, agriculture, distribution, production and traffic, labour statistics, etc.) together with descriptions of their main characteristics (e.g. timing, content, method, *published form*). Also very helpful and inexpensive are the six booklets in the *Guides to Official Sources* series, prepared by the Interdepartmental Committee on Social and Economic Research and issued by H.M.S.O.,

---

[1] The price of the main catalogue is £14.

London. These usually cover in detail statistics relating to the work of one particular Government Department (in brackets below), though there is cross-referencing to related material published by other Departments. The complete list, to date, is as follows:

1 (1958) *Labour Statistics* (Ministry of Labour—employment, unemployment, wages, earnings, hours, industrial relations)
2 (1951) *Census Reports of Great Britain, 1801–1931*
3 (1953) *Local Government Statistics* (Ministry of Housing and Local Government—rates, housing, water and sewerage)
4 (1958) *Agriculture and Food Statistics* (Ministry of Agriculture, Fisheries and Food)
5 (1961) *Social Security Statistics* (Ministry of Pensions and National Insurance and the National Assistance Board)
6 (1961) *Census of Production Reports*.

The fullest and longest established of all the sources of British statistics is, of course, the census, and an account of its contents will form a framework for much of the remainder of this section, supplemented by references to other sources which are relevant to a particular topic. A review of some of the more general aspects of the Census will precede more detailed description of the various parts, but in any extensive work using Census volumes the 'Guide to Official Sources, 2' mentioned above is invaluable. So too is B. Benjamin's *The Population Census* (London, 1970), No. 7 in the *Reviews of Current Research* series initiated by the Social Science Research Council and which has the advantage of describing material included in the 1961 and 1966 Censuses. With the exception of 1941, when there was no census, and 1966 when a new 'intermediate' census[1] was added, the British Census has been taken at ten-yearly intervals since 1801 by means of questionnaires completed by householders and returned to the Register-General for co-ordination and condensation. Because of the personal nature of much of their contents the actual questionnaires are only released for research work after a hundred years has elapsed, so that for the Censuses of 1881 and after only the published information, or certain abstractions of the unpublished information, (see below) are available. The multi-volume nature of the published results of censuses has already been hinted at and is illustrated by the fact that the Census for 1871 occupied eight volumes, that for 1911 seventeen and that for 1951 well over fifty so that it is essential, for library work, to know which volume or volumes one actually requires. Unfortunately, no consistent arrangement of material has been followed, though variations on three themes can be observed. There are:

1 presenting all the information on *one topic* for the whole country in a single volume
2 presenting all the information on all topics for *one county* in a single volume
3 presenting *most* of the information for one county in one volume, but omitting certain topics which are covered by volumes relating to the whole country.

The last-mentioned method is the one now in use. Since 1861 the Census of Scotland has been separate from, though coincidental with that for England and Wales, and the Isle of Man and the Channel Islands have had separate volumes since 1881.

[1] The 1966 intermediate Census was based *entirely* on a 10% sample of all households. For this reason the published results differ somewhat in format, scope and content from those of other censuses.

Publication of the results of the Census, hitherto a somewhat protracted process but now much accelerated by the use of computers for handling, sorting and producing the information, usually begins with a Preliminary Report, a most useful volume which includes an outline of essential information such as the population of each Local Authority and comparable figures for the previous census with the difference expressed as a percentage change. This has the advantage of presenting the most frequently sought-after information without a troublesome time-lag and within one volume covering the whole country.

At first sight the Census appears to offer the possibility of obtaining comparable information on many topics over extended periods during the last 150 years, but in practice this proves to be very difficult to achieve. Census information is often not directly comparable with that obtained ten years before, for a variety of reasons among which may be listed:

1   the increasing scope of the Census. Some topics are of only recent inclusion, e.g. workplace (journey to work), which appeared first in 1921, was omitted in 1931 and reincluded from 1951 onwards
2   changes in the nature of the subject, e.g. the number of employed persons which is affected by variations in school-leaving age
3   changes in definition, e.g. what constitutes a house or dwelling
4   changes in the *type* of areas for which information is given
5   changes in the *boundaries* of areas for which information is given.

To these one might add changes in the amount and detail of information given for certain areas. Not all areas are presented in the same detail. Rural Districts are only subdivided into parishes for totals of population and houses; towns of over 50,000 population have more detailed information than others, 5,000 population also forms a minor distinguishing point, more important in the past than today, whilst 15,000 population was the minimum size of authority for which information was published in the 1966 sample Census. Change in status, e.g. from rural parish to Urban District, will affect the amount of information available, and it is well to check the practicability of abstracting a long run of figures before starting to do so. Full details of most of the kinds of change described above are usually included in the introduction to the relevant Census volumes, but those listed under 4 and 5 may need some elaboration.

Today the modern administrative structure of counties, county boroughs and county districts (Municipal Boroughs, Urban and Rural Districts) underlies the presentation of information throughout the Census,[1] but since this hierarchy was essentially evolved only in the late nineteenth century, early censuses show, instead, figures for a bewildering variety of areas such as Ancient Counties, wapentakes, townships, etc. Figure 58 shows the evolution of the modern system from the ancient units and should assist in using the early volumes and in assessing comparability of statistics relating to different types of area. Unfortunately, the diagram can show changes *in general terms only*, since despite the attentions of Parliament anomalies and exceptions survive to defeat the establishment of a

[1] Figures for others areas, e.g. ecclesiastical parishes, are given, but the general arrangement is made under headings relating to contemporary Local Authorities.

uniform system of Local Government units. Thus two Acts of Parliament still left un-scathed in 1931 fourteen of the oddities known as extra-parochial places, and even in 1951 the Census was bound to admit that Lundy Island 'appears to be extra-parochial' or, of York Castle that 'its exact status is uncertain'; these we can regard as examples of a much-reduced host of anomalous units or 'illogical' areas which in the nineteenth century made such a complex tangle of boundaries in Britain.

To modern eyes two important characteristics differentiate the arrangement of the old units from that of the new. In the first place the old units were often not physical entities, but had many detached parts of all sizes often many miles from the parent area; secondly, each large unit rarely contained a *whole number* of small units. Examples of the first type survive in the well-known detached parts of Worcestershire and Flintshire, but this will scarcely prepare the uninitiated for parishes in twenty or more separate parts, or for a former detached portion of County Durham at Crayke, near York, quite twenty-five miles beyond the county boundary. Many of these 'parts' were removed under Acts of 1844 (counties) and 1882 (parishes) and fortunately in the latter case, where they were most numerous, they were often of land only, without population.

On the whole the second characteristic is more troublesome statistically. Apart from hundreds and wapentakes, which always seem to have fitted exactly into counties, there was no self-contingency of smaller units within the larger ones. Thus townships ran into two parishes, parishes ran into two counties, as also did Urban and Rural Sanitary Districts and particularly Poor Law Unions (Stamford Union extended into *five* counties), a state of affairs which was responsible for many of the differences between the ancient counties and the new administrative ones when these were created. The Local Government Act of 1894 had the last word on this subject, declaring that 'no Civil Parish could be partly in an Urban and partly in a Rural District or in two Administrative Counties'.

Another term appearing in Figure 58 and also likely to cause confusion is the word 'parish', which even today retains a dual meaning, 'ecclesiastical' often being implied in common usage, but 'civil' in most official documents. Originally of ecclesiastical origin and referring broadly to the area attached to a particular church, it maintained a parallel existence with the ancient unit of local territorial organisation in Saxon England—namely, the *township* (i.e. a settlement and the lands belonging to it). So long as each village had its own church the two areas would frequently coincide, and where this happened the term 'parish' tended to supplant the older one, as in much of southern England where parish and township once coincided. In the poorer northern counties, however, churches were fewer and *several* townships frequently formed a *single* parish, giving the appearance of a two-tier structure, though matters were not always so simple as this. There was the usual overlapping of boundaries, e.g. in the West Riding Cumberworth Half township extended into three parishes and two wapentakes, and there were also extra-parochial places, neither townships nor parishes and which had a strange variety of origins.

Enough has been said to show that throughout Britain under this system the smallest local units had neither common terminology nor common status and, therefore, in the Census Report of 1871 the term 'Civil Parish' was introduced to define the local unit of territorial organisation. The old 'simple' or undivided ecclesiastical parish became a civil parish, but so also did the old townships in 'compound' parishes and many of the

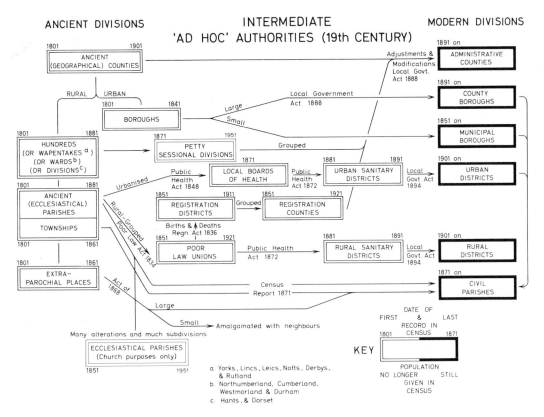

Figure 58 Relationship of ancient and modern administrative areas in England and Wales and their occurrence in census reports.

old extra-parochial places too if they were large enough.[1] Rather confusingly each larger unit such as an Urban District, Borough or County Borough was also made a Civil Parish and this remains the general rule today[2] so that the term 'Civil Parish' applies to units of all sizes.

The various Acts of Parliament which attempted to rationalise the boundaries of these *Civil* Parishes had no powers to alter those of the old *ecclesiastical* parishes, and the two units began to exist separately, side by side, but often relating to different areas. The separation persists to the present day, complicated by much subdivision of ecclesiastical parishes in urban areas, as new churches were built, and amalgamation in rural areas.

[1] By an Act of 1868. Small ones were supposed to be amalgamated with their neighbours, but quite a few 'escaped' (see above).

[2] In a few exceptional cases *several* civil parishes were formerly found within a larger unit, e.g. Malvern U.D. (Worcs.) contained 4 civil parishes; Much Wenlock M.B. (Salop.) 9, and the County Borough of Kingston-upon-Hull 2. This practice now appears to have ceased so that urban local government units now form entire civil parishes.

Figures for the populations of ecclesiastical parishes were included in Census Reports until 1951 but are little used, largely because there is no published map of ecclesiastical parish boundaries. This is rather unfortunate, especially in some cities and large towns where ecclesiastical parishes form smaller and often more 'local' units than the 'wards' for which the Census provides figures.[1]

Between the ancient and modern systems of local government units stood the need for many, often drastic, boundary changes so that it sometimes becomes necessary to establish the exact extent of areas for which figures are given at different dates. This may prove surprisingly difficult. The Census Reports have always given, in a separate table, details of all boundary changes since the previous Census, together with figures (for population and area only) to allow adjustment of either total to the other area, but they have never included a map showing the *extent* of these changes. For local boundary changes of almost any date enquiry at the Town Hall will provide an answer, but for areas further afield two sources may be used. The 'official' record of boundaries of all kinds is the 1:100,000 (formerly $\frac{1}{2}$ in to the mile) County Administrative Diagrams, published by the Ordnance Survey. These are frequently revised and are admirable for delineating modern boundaries, but older editions are not easily come by and former boundaries can often be more easily traced from the ordinary Ordnance Survey topographical maps. Of these the 1:2500 (25 in to the mile) is the map par excellence for boundaries, which are recorded in great detail; the 1:10,000/6 in map likewise shows very complete information but these large scales make it difficult to gain access to cover for extensive areas. Fortunately, all editions of the 'one inch' (1·6 cm to 1 km) map except the 'Popular' (or 4th Edition), current from 1920 to about 1939, have recorded parish boundaries, and since copies of old editions are often found in libraries or second-hand, these offer the most readily accessible solution to the problem. The parishes shown are usually 'ancient' (ecclesiastical) ones until about 1880, but civil parishes after that date; the approximate periods during which the various editions were current in England and Wales are listed below.

Edition	Current
1st 'Old Series'	1801 (south)–1870 (north)
2nd 'New Series'	1870–1885
3rd 'Small Sheet Series'	1885–1920
3rd 'Large Sheet Series'	1905–1920
5th	1930–1939 (south of about Birmingham only)

Despite the number and variety of these transforming boundary changes their main trends can be summarised fairly easily. In the late nineteenth century new urban and rural units had been created, either by adapting ancient ones or by carving out areas quite unrelated to older boundaries; the whole range had been 'tidied up' and the modern units had emerged in embryo by 1894. Since then five main types of change have occurred, namely:

1   piecemeal extensions to growing towns and cities
2   amalgamations of many rural parishes

[1] *Crockford's Clerical Directory* conveniently records the population of all ecclesiastical parishes (at the date of the last census).

3   amalgamations of many small urban districts
4   creation of some new urban districts
5   reorganisation of the *grouping* of units in rural areas into new Rural Districts.

Following the passing of the Local Government Act, 1929, particularly widespread and drastic boundary changes occurred, so much so that the 1931 Census, which described the areas before revision, had to be augmented by special tables (Part II of each County Volume) recording at least the population and area of each of the newly constituted authorities. Figure 59 illustrates an area where particularly marked changes separate the ancient and modern boundaries and helps to give some idea of the difficulties which may arise from this source. The impending reorganisation of local government in Britain will introduce yet another hiatus into the areas traditionally used for census purposes.

One obvious way of counteracting the worst reorganisation of this sort is to have figures for smaller areas than local authority areas so that these can then be regrouped to provide equivalents of new areas, past areas and so on, as well as incidentally providing welcome finer detail about an area generally. In this respect the *published* volumes of the Census are of little use[1] but it is possible to obtain from the General Register Office totals for *enumeration districts*, which are the basic areas used in collecting data for the Census and which contain on average about 200 households. Unfortunately a charge has to be made for supplying this information and it is perhaps worth pointing out that in many cases for the 1951, 1961, and 1966 Censuses it has already been purchased by local authorities particularly for planning purposes. Whilst the authorities cannot be expected to make the figures available to all comers experience has shown that they are often sympathetic to requests from *bona-fide* research workers.

In the 1971 and subsequent Censuses it is hoped that *geocoding* of information will enable totals to be supplied for *any* kind of area whose extent can be defined in terms of grid references, e.g. a 1 km or 100 metre square, though once again payment will be required comensurate with the labour involved in extracting the information.

Having introduced the census in general terms we are now in a position to review some of the main topics it describes and add further sources of information.

Population

A recording of the total population of various areas has always formed a primary aim of the Census; it is unfortunate that, obtained in this manner, statistics only become available every ten years (every 5 years if the practice of taking an intermediate census as in 1966 is continued). Populations recorded are *de facto* and not *de jure*, i.e. the population *actually present* on census night and not the population which *ought to have been or would normally have been present*. The difference is usually small[2] (see, for example, Census of England and Wales, 1961, *Usual Residence* volume), but can occasionally be troublesome. The unusual example of Leyburn has already been seen in Chapter 7, and the following two footnotes

---

[1] Figures are given for wards and civil parishes in published volumes but these refer to an extremely limited number of topics—generally population, dwellings and household totals only.

[2] Typical errors are up to $\pm 3\%$ but these may rise to $\pm 10\%$ or more in unusual cases.

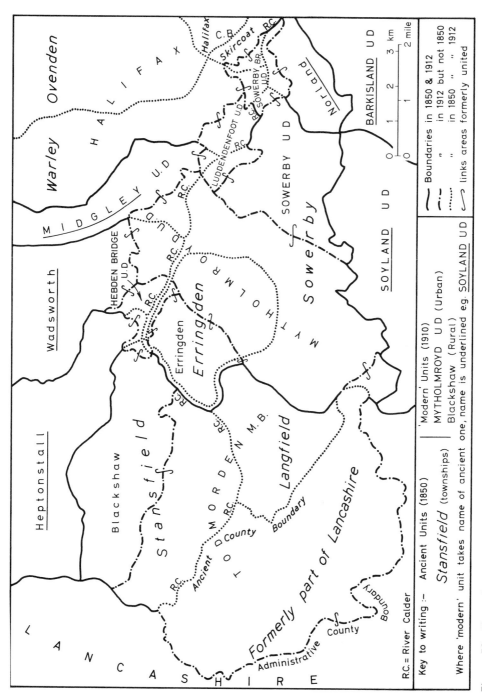

Figure 59 Upper Calder Valley (West Yorkshire)—administrative areas before and after evolution of 'modern' units in the late nineteenth century. Note the carving-out of valley-centred urban units (Todmorden M.B.; Hebden Bridge U.D.; Mytholmroyd U.D.; Luddendenfoot U.D. and Sowerby Bridge U.D.) from old plateau-centred townships. Changes made in the 1930s confused the position even more. Hebden Bridge U.D. and Mytholmroyd U.D. amalgamated to form Hebden Royd U.D.; Sowerby Bridge U.D., Luddendenfoot U.D., Norland parish and part of Midgley U.D. (the rest to Wadsworth parish) became Sowerby Bridge U.D. and Soyland U.D. and Barkisland U.D. became part of Ripponden U.D.

Source: O.S. Maps, 6 in. Yorkshire (1st Edition, 1850) and 1 in. (3rd Edition, 1912).

to the 1911 Census will emphasise the point:

> 'The population in 1911 of Lyme Regis Civil Parish includes 473 persons on board H.M. Ships.'

> 'The large increase in the population of Lancing Civil Parish is attributed mainly to the presence of students and staff of Lancing College, which was closed for the Easter vacation at the Census of 1901.'

The most troublesome discrepancies of this kind usually have their origins in variation in 'institutional population', which may be considerable (e.g. in 1961 over 18% of the population of Bodmin M. B. (Cornwall) were in psychiatric hospitals), and the best means of avoiding mistakes from this source is to use the figures for the population 'in private households', already referred to in Chapter 7. A very useful source for these figures together with basic totals for population, dwellings and households down to ward and rural parish level are the *Population, Dwellings and Household* pamphlets published for pairs or groups of counties in the 1961 Census (and presumably in subsequent Censuses). Priced at only a few pence each they provide a cheap and convenient source of basic data over a wide area.

Intercensal change receives considerable attention in Census Reports. Comparable figures for the same area at the previous census are provided and the change is usually recorded as a percentage and also subdivided into components of natural increase and migration.

For statistics of population at dates between censuses, only estimates are available, the fullest account being contained in the *Register-General's Statistical Review of England and Wales*[1] which records not only estimated population for local authority areas but also such features as births, deaths, birth rates, death rates, marriages, and local government elections. Publication is normally in three volumes, Part I, Medical (relating mainly to mortality), Part II, Population, and Part III, Commentary, but the much cheaper pamphlet *The Registrar General's Annual Estimate of the Population of England and Wales and of Local Authority Areas* (price about 15p) provides a useful summary of the actual population totals.

Among the considerable amount of information on other aspects of population found in the census reports are tables covering age and sex distribution among local populations,[2] socio-economic groups, marital condition and birthplace, as well as details of the possession of various household arrangements such as number of rooms, piped water or a bath. Since 1966 educational history, car-ownership and garaging facilities have been added to the traditional list.

Among statistical sources censuses are unusual in that they are able to answer not only 'How much or how many', but also to go a considerable way towards providing the answer

---

[1] Published by H.M.S.O., London, as are all Government publications referred to in this chapter unless otherwise specified. There is a comparison volume for Scotland.

[2] The author very regretfully concludes that he can find no foundation for the statement, which appeared in a recently published humorous article, that these volumes include a table headed 'Population of Great Britain, broken down by age and sex'.

to 'Where?' as well. For more precise information on the location of population the best source is the annual *Register of Electors* which provides, for each parish or polling district the names *and addresses* of all adults entitled to vote at elections. As such it can be used to establish a very detailed distribution of the *adult* population, but unfortunately no easy way exists of extending this to cover *total* population. The usual practice is to obtain an estimate of total population in the area covered by the register and allot a proportional number of children to each adult; this is often reasonably accurate for large blocks of adult population but cannot be expected to work too well for small groups. Copies of the Register can be obtained at a small fee from the appropriate Electoral Registration Officer (usually the Clerk to the County or County Borough Council) who is also responsible for retaining previous Registers, from which historical comparisons can be made and changes determined.

Industry and occupation
A second great preoccupation of the Census has been to establish some account of the occupations of the population. The problem has proved enormously complicated and early results in particular were not very successful or reliable. The great difficulty was to establish a suitable classification so that people who described themselves as 'sad iron grinders' (engineering), 'bottom knockers' (pottery), 'banknote finishers' (printing) or 'twisters' (many industries), along with thousands of other categories could somehow be accommodated within a classification of manageable proportions. At first a relatively small number of extremely broad groups was used, but from 1851 onwards there began to evolve the classification of occupations by classes and sub-classes, later called orders and sub-orders, which has been maintained with constant modifications to the present day.

Occupational statistics, it should be stressed, relate to the *kind of work a person does, irrespective of the industry which employs him*, and the twenty-seven main orders contain therefore such groups as 'clerks and typists', 'painters and decorators', 'warehousemen' and 'workers in unskilled occupations', many of whom can and often do move about from one industry to another. In this respect they should be carefully distinguished from a second set of figures, the '*Industry* Tables' which have appeared in the Census since 1921 and in which a worker is classified by the *industry which employs him, irrespective of what job he does*. The distinction needs watching, particularly since many of the headings used in the two tables are not greatly different from one another at first sight. Nevertheless, the distinction is quite clear cut. So far as the Occupation Tables go a lorry driver, for example, falls unambiguously into Order XIX, 'Persons employed in Transport and Communications'; from the Industrial Table's point of view he might be allocated to any one of the main industrial groups, to 'Transport and Communications', to 'Distributive Trades' and so on, according to the main business of his employer. The more 'industrial' terms in the Occupation Tables usually refer only to workers actually making the product of the industry and not to ancillary staff such as office workers, managers, cleaners, etc., so that the total for a particular industry in the Industry Tables will usually be considerably larger than that for the most comparable heading in the Occupation Tables.

One more distinction must be made. 'Occupation' is a *personal* characteristic and the tables have therefore always been arranged to show what *occupations the people of X follow, no matter where these take place*. The Industry Tables for 1931 followed a similar arrangement and show *the industries which occupied the people of X no matter where they were*. This was different from both 1921, 1951 and subsequently where the Industry Tables had a *place* type of arrangement showing totals for *the numbers employed in the industries in X no matter where the workers came from*, so that comparisons with the 1931 figures cannot be made. The most striking example of the differences involved comes from the figures for the City of London shown below:

	Census of			
	1921	*1931*	*1951*	*1961*
Occupation Tables, total occupied	8,996	7,390*	3,591	3,440
Industry Tables, total employed	†	6,978*	337,486	390,730

* Theoretically these two figures should be the same; the reasons for the difference are not given.
† Separate figures were not provided for units with small populations at the 1921 census.

The table for Wigan (Lancashire), on p. 189, illustrates the kind of differences which can occur in a medium-sized town. Note, for example, that the Industry Tables 'played down' the role of mining, simply because most of Wigan's miners worked at collieries outside the Borough Boundary. Unfortunately details of Occupation and Industry were among the several topics which were asked of only a 10% sample of households in the 1961 Census and for this reason the *published* figures for England and Wales were restricted to those for areas with over 50,000 population[1] instead of all local authorities as formerly. The loss of this long established and basic published data for smaller towns[2] was greatly to be regretted and much reduced the usefulness of the Occupation, Industry and Socio-Economic Group pamphlets (similar to the Population, Dwellings and Households pamphlets referred to above) which were introduced with the 1961 Census.

As with population this decennial census information often needs supplementing in between censuses, but it is difficult to get anything approaching the detailed *local* figures which the Census gives. Regular figures on manpower, and such other topics as unemployment, short time and overtime and labour turnover are collected by the Ministry of Labour and published in the *Ministry of Labour Gazette*, issued monthly. Totals are usually for the whole country, subdivided by industry in the same manner as the Census Industry Tables;[3] the only local figures given are for unemployment (each month), but even these relate to Employment Exchange Areas which do not coincide with, or often even approximate to, the Local Authority Areas used for census purposes. The annual exchange of Employment Cards at each Employment Exchange makes it possible, once a year, to establish the way in which workers registered at that exchange are subdivided amongst the various industries. Unfortunately, these figures are not published, but from the frequency with which they appear in theses, town planning analyses, etc., it would appear that they are often disclosed to *bona fide* research workers.

[1] 10,000 in the case of Scotland.
[2] Presumably the *unpublished* totals for individual towns can be obtained from the General Register Office.
[3] Following the drawing up of a Standard Industrial Classification in 1948 the same subdivisions have been used, where applicable, in the Industry Tables of the Census and for figures prepared by the Ministry of Labour and the Board of Trade.

For some industries local information may be obtained from a yearbook or similar sources, some of which are excellent, e.g. the *Colliery Year Book and Coal Trades Directory* which until 1963 listed men employed above and below ground and annual production at each colliery in Britain and many in Western Europe and the U.S.A.;[1] in addition most of the nationalised industries and organisations publish an 'Annual Report' or similarly titled volume dealing with their activities and containing varying amounts of statistics.

Unlike the Census, yearbooks often assess industry from a *production* rather than a *manpower* point of view and for wider information of this kind quite a different set of official sources can be consulted. The fullest of these are the Censuses of Production which have been taken by the Board of Trade in 1907, 1912, 1924, 1930, 1935, 1946 (partial), 1948 and in various forms yearly since then, although more fully at three-to-five-yearly intervals from 1948. As well as details of production both by quantity and value the censuses have recorded employment, number and size of establishments, costs, investment, fuel and power equipment, analysed mainly by industries for the whole of Great Britain, but recently with increasing breakdown into smaller units such as countries and Standard Regions.[2] Publication is in multivolume form, often summary volumes plus several sections devoted to groups of the various industries covered by the census, i.e. broadly those in orders II to XVIII of the Standard Industrial Classification although this has varied slightly in the past.

Parallel in many ways to the Census of Production but dealing with retail distribution and a very few of the service trades are the *Censuses of Distribution* taken by the Board of Trade in 1950, 1957 (sample), 1961 and 1966 (sample) and which give figures for number of establishments, wages, turnover and capital expenditure for different sectors of distributional activity and for Counties, Standard Regions, Conurbations and individual local authority areas. Detail for particular towns varies a good deal according to size and type of Census, being omitted from sample Censuses: 2,500, 20,000 and 200,000 population have been used as important size differentiators but for the largest size group information is very full, distinguishing even between 'city centre' and 'suburban' totals.

Locating the units of industrial production or retail distribution is, perhaps, rather easier than trying to 'tie down' the population. As a first source of information the *Classified Telephone Directory* proves most useful, whilst many trades and industries have their own yearbook or directory. Once again the Ordnance Survey maps can be a great help, the largest scale map published for the principle urban areas at 1 : 1250 (50 in to a mile) showing the use of all buildings which are not dwelling houses, offices or shops. The 1 : 2500 (25 in) and 1 : 10,000 (6 in to the mile) maps do the same at least for many of the

[1] It has since been superceded by the *Guide to the Coalfields* published at 3-yearly intervals since 1963 and dealing with Great Britain only. *The Municipal Yearbook* is another useful source of statistics relating to Local Authority activities of all kinds.

[2] The Standard Regions have been widely used in post-war official statistical publications where a breakdown of the country into medium-sized areas is required. Though care is needed in comparisons over time as there have been changes in numbers (nine to eight in England) and boundaries whilst often maintaining similarities of title. The constituents of each region will usually be found specified in any of the national Census volumes which make reference to them. The Standard Conurbations, which originated in the 1951 Census are yet another type of national region for which figures are often provided in official publications.

Wigan (Lancs.) Census 1951

Occupational Tables			*Industry Tables*	
Total occupied persons	41,351	35,255	Total employed, all industries	
Order			*Order*	
I Fishermen	1			
II Agricultural, etc., occupations	230	110	Agriculture, forestry, fishing	I
III Mining and quarrying occupations	5,194	438	Mining and quarrying	II
IV Workers in glass, ceramics, Cement, etc.	72	123	Ceramics, glass, cement	III
V Coal gas makers, workers in chemicals	328	421	Chemicals and allied trades	IV
VI Workers in metal manufacture, engineering	3,815	167	Metal manufacture	V
		2,387	Engineering, shipbuilding, Electrical goods	VI
VII Textile workers	4,055	585	Vehicles	VII
VIII Leather workers, fur dressers	448	759	Metal Goods (n.e.s.*)	VIII
IX Makers of Textile Goods and articles of dress	2,301	39	Precision instruments and Jewellery	IX
X Makers of food, drink and tobacco	543	6,368	Textiles	X
		163	Leather goods and fur	XI
XI Workers in wood, cane or cork	966			
XII Workers in paper, printers	188	3,333	Clothing	XII
XIII Makers of other products	264	986	Food, drink, tobacco	XIII
		692	Manufacture of wood and cork	XIV
XIV Workers in Building and Contracting	1,851	348	Paper and printing	XV
XV Painters and decorators	455	182	Other manufacturing industries	XVI
XVI Administrators, directors, managers (n.e.s.*)	467	1,900	Building and contracting	XVII
XVII Persons employed in transport	2,541	684	Gas, water, electricity	XVIII
XVIII Commercial, finance, etc. (excl. Clerical)	4,038	2,757	Transport and communications	XIX
XIX Professional and technical (excl. Clerical)	1,677	5,872	Distributive trades	XX
		655	Insurance, banking, finance	XXI
XX Persons employed in defence services	408	2,076	Professional services	XXII
XXI Persons engaged in entertainment and sport	131	1,679	Public administration and defence	XXIII
XXII Persons engaged in personal service	2,543	2,503	Miscellaneous services	XXIV
XXIII Clerks, typists, etc.	2,325			
XXIV Warehousemen, storekeepers, packers, etc.	869			
XXV Stationary engine drivers, stokers, etc.	541	18	Not stated, ill-defined	
XXVI Workers in unskilled occupations (n.e.s.*)	4,835			
XXVII Others and undefined	275			
XXVIII Retired, not gainfully occupied	23,849			
Total (equal to population aged 15 and over)	65,200			

*not elsewhere specified

larger units, and for big industrial premises usually give the name of the works and the industry carried on there; older editions of these maps can sometimes be used to obtain former distributions and check changes of use, but it should be noted that the names of the premises in particular are apt to change (the same applies to farms shown on O.S. maps), so that a map several years old may not be a great help in locating establishments listed by their present names. Formerly even 1 in to the mile maps marked the nature of many sizeable isolated industrial premises, e.g. Mill (Cotton), Colliery, Mine (Barytes), and so on, but modern practice seems to be to reduce this to the much less specific 'works' or 'mine'.

Agriculture

Though details of agricultural employment are recorded in the Census it is necessary, as with industry, to look elsewhere for details of production. Fortunately, quite complete information is available, arising from an annual return made each June to the Ministry of Agriculture from all holdings of more than one acre. The principal statistical results from this return are published annually by the Ministry in *Agricultural Statistics—England and Wales.*[1] Information provided includes acreage of land under a wide variety of crops, grass and rough grazings, totals of various livestock, estimated yields of crops, size of holdings and workers employed, and fortunately for a great deal of this information *county* totals are available as well as national ones. A second section on horticulture adds figures for vegetables and fruit in rather greater detail.

Local information for areas smaller than counties is not published but the Ministry are prepared to extract totals under most of the headings listed above for any particular parish, series of parishes or for National Agricultural Advisory Service Districts which are amalgamations of parishes. An estimate of the approximate cost[2] of this work will be provided on application to Block 'B', Agricultural Census Branch, Government Buildings, Epsom Road, Guildford, Surrey, although research workers who are able to visit Guildford are allowed, subject to prior arrangements, to extract the totals themselves, free of charge. The annual nature of the census means that runs of figures are quite easily obtained and the results from earlier censuses allow ample scope for detection of changes.

A record of the distribution of the six main types of land use in Britain is provided by the maps published at 1·6 cm to 1 km (1 in to a mile) to illustrate the Land Utilisation Survey of Britain, carried out in 1931 (England) and 1932 (Wales and Scotland). Unfortunately there was a very considerable delay in publishing many of these sheets and it should be remembered when using them that whatever the date of publication the information shown relates to the period of the survey and not later. Accompanying the maps was a report of the Survey, published in 92 parts (one for each administrative county in Britain) under the title of *The Land of Britain*, whilst a later work, L. D. Stamp *The Land of Britain, Its Use and Misuse* (third edition), London, 1962, provides a co-ordinating account of the survey generally and analyses its results. Both of these accounts, and the county volumes in particular, form excellent supplementary sources to the maps, providing an enormous amount of information on many local distributions besides those of land

[1] There is a similar volume for Scotland.

[2] The present basis of the charge is one new penny per item.

use. The maps show the six main categories of forest and woodland; meadowland and permanent grass; arable; heathland, moorland and rough pasture; houses with gardens, orchards and nurseries; and land agriculturally unproductive. Unfortunately, many are now out of print and those that remain, as well as the county volumes, are obtainable only through Messrs. Edward Stanford, Ltd., 12–14 Long Acre, London, W.C.2.

A similar, more ambitious project, the Second Land Utilisation Survey of Britain has recently been completed under the direction of Miss Alice Coleman, M.A., King's College, London. Although the whole survey has been taken and maps exist in manuscript form the published sheets, which are at 1 :25000 scale ($2\frac{1}{2}$ in to 1 mile) number only about a hundred of the several hundred which will ultimately be needed. The material recorded is in much greater detail than on the earlier survey, several varieties of cropland being shown as well as useful information such as the type of industry carried on at the various industrial premises. A series of county reports is contemplated, but none is yet published.

These surveys apart, it is worth while remembering that most Ordnance Survey maps, even down to the one inch scale, have always differentiated important types of land use such as woodland, moorland and rough pasture, marsh, orchards and (more recently) glasshouses.

Transport and communications
It is an unrewarding task to try to obtain details of *local* movements of people, goods or vehicles in Britain. For railways, fairly detailed information, though in units of oddly varying sizes, was formerly available in the annual 'Railway Returns',[1] but the amalgamation of the railways into four large main line companies in 1923 removed most of the local element from this source and figures became available generally only for large sprawling entities without geographical coherence. The only exceptions to this rule were the few lines which were not amalgamated (usually impecunious local oddities) or lines jointly owned or worked, for which figures continued to be provided until the Railway Returns ceased in 1938; these exceptions occasionally contain a few surprisingly useful local systems, such as the Liverpool Overhead Railway, the Mersey Railway and the Manchester South Junction and Altrincham line. Information is given on such items as capital, receipts, working expenses and traffic.

Nationalisation did not remove this general drawback of arrangement by very large regions (admittedly more geographically compact), and figures for these appeared for a time in *Transport Statistics* published monthly by the British Transport Commission. Current publication which provides a division only into British Railways and London Transport is in *Passenger Transport in Great Britain* published annually by the Ministry of Transport and which also gives figures for passenger traffic by road and air.

Details of the number of trains using sections of railway can be obtained from the public timetables (passenger trains) or the less-well-known Working Timetables (all trains), although the latter source is not on sale to the general public.

For information on local road traffic, though not on amounts of goods or passengers carried, the position is a little better. Traffic censuses have been taken by the Ministry of

[1] Issued annually by the Ministry of Transport.

Transport since 1922, usually at three- or four-yearly intervals up to 1938, 1950, 1954, 1961, 1965, 1969 and annually since then. Most of these relate to Trunk Roads only and unfortunately even where the results of these censuses have been published, the Reports deal only with general aspects and do not give figures for individual census points. Local information of this sort is usually best obtained through the appropriate Highway Authority (usually the County or County Borough Council) who in fact were responsible for obtaining the information for the Ministry on the occasion of each census.

Figures relating to road transport for the whole country will be found in *Passenger Transport in Great Britain* referred to above; the numbers of motor vehicles licensed for each County and County Borough is given in *Highway Statistics* issued annually by the Ministry of Transport.

A useful account of local *personal* movements is contained in the 'Journey to Work' tables found in the '*Workplace*' volumes for the 1921, 1951, 1961 and 1966 Censuses. Movements are classified by Local Authority areas and the data recorded form a useful basis for a study of commuting Britain.

Traffic at the various British ports, along with an enormously detailed account of British overseas trade generally will be found in the four- or five-volume *Annual Statement of Trade of the United Kingdom*, or in the annual *Digest of Port Statistics*, published by the National Ports council.

In any search for sources of statistical information on all aspects of life in Britain government publications must of necessity provide the majority of the results. As all library users will know, few types of publication are apt to cause so much trouble in indexing or in obtaining particular titles and 'authorship'; for obtaining specific information of this sort as well as a comprehensive account of publications of all kinds works such as *Catalogue of Parliamentary Papers, 1801–1900; General Index to Parliamentary Papers, 1900–1949* and the annual *Catalogue of Government Publications* will be found invaluable. The last-named, in addition to a full list of British official publications has for many years listed those of International Organisations including the Statistical Office of the European Communities and the United Nations.

General index

Index of places referred to in the text or diagrams